Silk, Spices, and Glory

Silk, Spices, and Glory

In Search of the Northwest Passage

M. A. Macpherson

Oct 19/09.

To Alex –

Explore the writing world.

Best Wishes – [illegible]

Front cover image of Arctic explorers from *Frank Leslie's Illustrated Newspaper*, December 1855, courtesy Russell A. Potter

Interior illustration by Grace Buzik
Cover design by Thinkinc Communications Ltd.
Interior design and map by Articulate Eye

The publisher gratefully acknowledges the support of The Canada Council for the Arts and the Department of Canadian Heritage.

We acknowledge the financial support of the Government of Canada through the Book Publishing Industry Development Program for our publishing activities.

Printed in Canada.

01 02 03 04 05/ 5 4 3 2 1

CANADIAN CATALOGUING IN PUBLICATION DATA

Macpherson, M. A. (Margaret A.), 1959–

Silk, spices, and glory

Includes bibliographical references.
ISBN 1-894004-52-3

1. Northwest Passage—Discovery and exploration. 2. Arctic regions—Discovery and exploration. 3. Explorers—Northwest Passage.
I. Title.
FC3961.2.M32 2001 910'.9163'27 C00-911410-6
G640.M32 2001

Published in Canada by
Fifth House Ltd.
A Fitzhenry & Whiteside Company
1511-1800 4 Street SW
Calgary, Alberta, Canada
T2S 2S5

Published in the U.S. by
Fitzhenry & Whiteside
121 Harvard Ave.
Suite 2
Allston, Massachusetts
01234

For KMS, who saw for herself.

And, always, for Mark.

Table of Contents

ARCTIC OCEAN
PARRY ISLANDS
M'CLURE STRAIT
MELVILLE ISLAND
CORN-WALLIS IS.
BARROW STR.
C. PROVIDENCE
BANKS ISLAND
VISCOUNT MELVILLE SOUND
PRINCE OF WALES STR.
PRINCE OF WALES ISLAND
PEEL SOUND
AMUNDSEN GULF
VICTORIA ISLAND
PRINCE ALBERT SD.
FRANKLIN STR.
VICTORIA STR.
KING WILLIAM ISLAND
QUEEN MAUD GULF
BACK
THE CANADIAN – ARCTIC –

SMITH SOUND
GREENLAND
BAFFIN BAY
LANCASTER SOUND
DAVIS STRAIT
BAFFIN ISLAND
FURY AND HECLA STR.
GULF OF BOOTHIA
CUMBERLAND SOUND
FOXE BASIN
FROBISHER BAY
PRINCE CHARLES IS.
RESOLUTION IS.
FOXE CHANNEL
ROES WELCOME SOUND
HUDSON STRAIT
NOTTINGHAM IS.
DIGGES IS.
UNGAVA BAY
BAY OF GODS MERCY
HUDSON BAY
Churchill
JAMES BAY

Introduction

The weight of gold in the palm, the touch of silk upon the skin, the tantalizing aroma of exotic spices: it was these things that first lured men into the icy waters at the top of the world. It was these things that drove them from hearth and home to pit themselves against nature's extremes.

The exploration of the islands and channels of the Canadian Arctic by Europeans makes for tales of foolhardiness and bravery, pride and despair, treachery and forbearance. It brought out the worst in human folly and the best of the human spirit. But more than anything else, the charting of the Northwest Passage was inspired by greed for the coveted commodities of the time: gold, silk, and spices.

It was the prospect of these things that danced before the mind's eyes of sailors when the ice fog enveloped their ships so they couldn't see the ice floes that threatened to destroy them, or when the frost on their beards was thick enough to hide the teeth loosening in their gums from scurvy, or when fear was the only thing they could smell in the crackling cold air of an Arctic night.

In a process that spanned five centuries, it was the dreams of spices, silk, and gold that sustained the men who sought them. Gaining access to these treasures would make them instant heroes and wealthy beyond their dreams. And the way to the treasures was always just beyond the next island, just past the next towering cliff of ice, just a day's sail away, or perhaps two, or three. But then, at last, the world of the East would reveal itself, and they would return triumphant to their countries and their monarchs, proclaiming that they had found the way, the Northwest Passage, the channel to riches untold.

While the Inuit and other aboriginal groups have explored the Arctic for millenia, the recorded European charting of navigable water west of Greenland began in the tenth century AD, when Norse explorers and coastal-dwelling peoples first braved the Arctic seas. Reports of the forests and fish of the ancient land later known as America are recorded from this period, and archaeological sites in southwestern Greenland and northern Labrador indicate that the great seafaring race known as the Vikings touched on these shores more than a thousand years ago.

The beginnings of truly serious and continuous European expeditions into the polar regions of the world date only from the late 1400s when the Anglo-Italian navigator John Cabot whispered three words—silk, spices, gold—into the ear of the English King Henry VII in 1496. Cabot's scheme was to find a quick passage to Asia via the northern latitudes. A few years before, Christopher Columbus had discovered what he thought were the fabled Spice Islands of the East and claimed them for Spain. Cabot knew England longed for strong trading links with the East, for the kingdom suffered as it watched Portuguese caravels and Spanish fleets carrying off the bounty of the Eastern markets. Ivory, silks, gold, precious stones, and, of course, spices were coveted in Europe. Britain was being left behind, and the king knew it.

Given vessels, Cabot reasoned, he could sail westward by the northern route and thus cut some two thousand leagues of dangerous sea travel off the course that currently saw sailors navigating the angry waters off the Cape of Good Hope at Africa's southern tip and battling fearsome pirates in order to make their way to the Far East. The English would be fastest and first, he claimed, and the world of trade would change forever once it was penetrated from the northwest.

John Cabot had the full attention of the English King, who called his cartographers to measure the world. Cabot's calculations, they found, proved to be true. But, like most men of his time, Cabot's geography was skewed. While all educated people knew that the world was round, few believed in the existence of America. There might be some islands (large islands, indeed!) standing in the way of the clear channel that would secure Eastern trade for England, but surely they were nothing a well-built ship and a sturdy crew could not circumnavigate.

Cabot was granted authority by the trade-hungry monarch, and eighteen men boarded the *Matthew* at Bristol on May 2, 1497. They sailed due west for the as-yet unknown yet surely easily accessed channel. Cabot claimed Newfoundland for the Crown on June 24 of that year. No doubt he and his men marvelled at the forests of tall timber that marched up to the shores and the abundance of fish in the sea. But this rocky, unknown land was obviously no Asian island marking the entrance to the Far East.

Disheartened but undefeated, Cabot returned to Henry, begging an opportunity to try again. This time equipped with six ships, Cabot and his son Sebastian were commissioned to explore the foreign coastline until they reached the Chinese Island of Cipano, famed for exotic spices. Thus was the eastern coast of North America first mapped and explored by Europeans. The consensus of opinion among noblemen and scholars in England was that this massive territory was not an island but a narrow continent, pierced somewhere in the middle to allow passage to the land of the great Khan. The dream would not die easily.

Throughout the early sixteenth century the idea of the Northwest Passage persisted, and much of the charting of Maritime Canada, the French settlements along the St. Lawrence River as far as Quebec, and the Labrador coast are the result, in part, of explorers like John Cabot and Jacques Cartier driving ever westward in search of the East. Explorers were intent on pushing through the continent of America on a waterway that would open eventually, inevitably, to Cathay.

While European cartographers were reluctantly altering and redrawing their maps as the extent of the "New World" became known, rumours of the Northwest Passage persisted, and the prices for silk and spices were a continued incentive to keep searching for them.

Some map-makers reasoned that the icy polar region must come to a peak at the top of the globe, as Africa and South

America did at the bottom. Expeditions should go farther north, following the land mass to its northernmost point. Once there, it would be a simple matter to sail around the top and down into the warm and tranquil waters of the Pacific.

The world these men encountered when they at last turned their prows farther north was one of unimaginable cold, of glaciated mountains, of land scraped raw of vegetation by the perpetual wind, and an endless night brightened only by the moon and stars, and the ghostly, surreal dance of the aurora borealis.

And ice. Everywhere, there was ice.

The polar ocean is, in fact, a continent of perpetual ice sitting squarely atop the world, with long, frozen arms descending south like the spokes of a wheel. But the polar ice cap is not a stationary mass; it rotates like the global wheel it resembles. Slowly, almost immeasurably, it is pushed by the currents of the Arctic Ocean first one way, then another, on the top of the spinning planet. Icebergs ceaselessly carve themselves off and fall from the mass, and the frozen spokes of the wheel are constantly shifting and changing. A solid path of ice in one year might be a channel of open water in the next. It is a world of infinite variation and alteration. The only constants are the perpetual ice itself, and air so cold and dry that it blisters exposed skin.

Pack ice, the bane of many an explorer, is formed by the annual freeze at the outer limits of the giant central glacier. It is a churning mass of ice and slush. It surrounds the permanently frozen cap and fills in the areas between the polar hub and its spokes. Cast about by tides, currents, and winds, the ice pack is looser and faster moving than the polar cap. Drifting bergs draw congealed slush to themselves, creating new ice, enlarging their mass until enormous islands are formed, islands capable of crushing a ship in an instant against the Arctic mainland or one of hundreds of granite islands that litter the archipelago.

The Arctic, we now know, is a labyrinth of channels and fjords, islands and peninsulas, and bays of such breadth as to rival the ocean itself. But early explorers did not know the shape of the world beneath the ice. Their journeys were forays into the unknown, to a land in which animals of mythic size roamed freely, in which human life was present only in the form of small bands of hunters dressed in the skins of animals that might as easily have killed them than been killed, in which darkness, ice, sickness, and cold prevailed in a sailing season that averaged six weeks out of the year.

Yet they went. From the discoveries of Cabot and the dreams of Eastern commerce, men of adventure and endurance continued to seek the fabled passage that joins the Atlantic to the Pacific. Each man added his piece to the puzzle. Each explorer extended the map. Many paid a terrible price. But ultimately the dream, born in late mediaeval England and nurtured for more than five hundred years, bore fruit in the discovery of the Northwest Passage.

This is the story of the men who made the dream come true.

MARTIN FROBISHER

Uncharted Waters

Martin Frobisher was a pirate, a scoundrel, and a crook. He was also the man who first turned his sights to the unexplored regions of the northwest after John Cabot's son Sebastian had downed his crude instruments and retired into quiet obscurity on the British coast and Jacques Cartier had returned to France after leaving his mark on the newly colonized Quebec.

Frobisher, a swashbuckling explorer with bull-headed determination, dared to go north into uncharted waters. By June 19, 1576, this determination had somehow translated itself into a two-vessel flotilla bearing the English flag and making its way north of the Shetland Islands, with Frobisher himself at the helm of one of the vessels. Ten years before, facing charges of piracy on the Guinea coast, he had faced down the Privy Council itself, pleading overzealousness for Queen and country. Branded a pirate, Frobisher lay low on home turf until he found a man willing to part with the capital that would enable Frobisher to sail away on another adventure. It is likely that he had been fleeing yet another impending unpleasantness on *terra firma* when, a year before, he managed to do just that, wresting funds from the merchants of London to finance his impossible journey to find the Northwest Passage. Frobisher might well have thought himself unstoppable.

Sending out a man with the reputation of a rogue and a record of indiscriminate looting was perhaps not in the best diplomatic interests of England, but it was, at the time, the island nation's best chance to limit the expanding powers of commerce and conquest of the Spanish kingdom to the south. Elizabeth I, having gained the throne eighteen years earlier, had subsequently watched the coffers of her empire dwindle while those of her southern neighbours overflowed. At the time Frobisher sailed, for example, the Straits of Magellan were virtually owned by the Portuguese and the Spanish. Some historians have likened the South Atlantic in 1575 to a large Spanish lake, so firmly were the eastern trade routes in the hands of the Spaniards and their cousins. And not only was the Spanish kingdom claiming sovereignty over much of the North American continent in 1576, Spain was also pressing dangerously close to the English coastline, with Spanish fleets drawing nearer to the Netherlands with every passing day.

Britain had little to lose, so when an irreverent Yorkshire mariner known as a pirate and a rogue proposed a mad dash for the Northwest Passage, it seemed a possible trump card in the British hand. They played Frobisher as a last bet. Disregarding his unsavoury past, they called upon him to carry the English flag through the Northwest Passage to the Indies and beyond.

The country had little real hope of finding a new trade route. But it was their last chance to thwart the Spanish. Better a pathetic attempt than no attempt at all, they reasoned, and so the Earl of Warwick petitioned the Crown to allow the eager Frobisher his dream. Immediately the ex-pirate began the process of soliciting money, ships, and equipment.

He went, hat in hand, to Michael Lok, a London shipbuilder and a gentleman who believed in the fabled Passage to the East. With influence and some ready cash, Lok helped Frobisher raise the money he needed to execute what the courtiers and capitalists in the city dubbed a "scientific voyage" of exploration. And while Frobisher's bravado far outweighed his skills as a navigator, he was glad of the chance to command a legitimate expedition. Fitted with two small ships, the *Michael* and the *Gabriel*, each around twenty-five tons, and an ill-fated, eight-oared pinnace, Frobisher and his motley crew of thirty-five set off on their voyage, armed with compasses, sundials, sand glasses, books on navigation, and inaccurate maps full of blank, formidable spaces that Captain Frobisher and his mate George Beste were meant to fill.

Not one to miss an opportunity for grandstanding, Frobisher is said to have left England via Greenwich Palace, Queen Elizabeth's temporary abode. With flags flying, ensigns flapping, and the overcoats of sailors strung up on the shrouds, he used his grand departure as a means of garnering royal support. But when the Queen sent a member of her court to bid farewell to the sailors, it wasn't enough for Frobisher. With the flotilla at anchor outside the palace, he rowed himself ashore and presented himself on bended knee as the faithful and fearless leader of the expedition that would find the Northwest Passage to Cathay in the name of Her Majesty and the sacred shores of England.

Something about Frobisher's enthusiastic, albeit unorthodox, behaviour must have struck a chord in the Queen, for she bade him Godspeed and sent him on his way to the polar lands under the powerful seal of a royal blessing. Rising from the position of common thief to royal commander in less than a decade, Martin Frobisher set forth for the Northwest Passage determined to do his country proud.

For eleven days the two vessels—the *Gabriel* piloted by Frobisher and the *Michael* by Captain Owen Gryffyn—forged west by northwest. And then a gale seemed to blow out of nowhere, and it buffeted the vessels like matchsticks in a whirlpool for eight full days. When the winds and seas finally abated and the weary men struggled to the relative calm of the foredecks, Frobisher and his entourage found themselves reduced by one vessel and four crew: the pinnace, similar to a modern lifeboat, had been swamped in the storm, her crew drowned in the brutal waters off the coast of Greenland.

Greenland was anything but a welcome sight to the mariners. Coupled with the loss of life so early in the voyage, the sight of this foreign landscape, a world of snow and ice cut like crystal, was enough to cause Owen Gryffyn to quit while fortune still smiled upon his little band of adventurers. It was evident, at least to the men aboard the *Michael,* that only more of the same awaited them should they continue into the unknown seas, and Master Gryffyn steered his vessel back on the course she had come. Not for him the cliffs of ice with their tops lost in the clouds. Not for him the monstrous ice floes ghosting by the silent ship. The *Michael* set course for home while Frobisher, his blood surging at the display of cowardice, prepared a launch and attempted a landing on the foreign shore. The attempt was futile, and who knows how many of the men aboard the *Gabriel* wished themselves on the *Michael* as they watched their sister ship run for home. Reduced by half only a month into their voyage, the men hauled their captain back aboard after his abortive attempt at landing amid the towers of ice. The haunting calls of seabirds pierced the fog. The strange, disembodied voices must have etched fear into the hearts of those who were destined to continue north under the command of Frobisher.

On July 13, pressing on around the southern tip of Greenland—then called Friesland by the Vikings—they ran into a second storm so violent that it made the first seem no more than a strong breeze. The *Gabriel* was beaten and cast on her side by the giant polar waves, which poured in through the open hatches on the main deck. The ship was filling with water and sure to sink. As the crew clung to the gunwales and prayed for a quick and easy death, Martin Frobisher sprang into action.

Grabbing a knife and dropping down the side of the ship, so bowed over that its rails were submerged, he cut through the stays that secured the mizzen-mast. Quick to see their master's logic—lightening the ship's windage—the sailors took axes and knives and hacked away at the mast until it pitched over into the dark churning sea. It seemed to be enough. With a mighty groan, the *Gabriel* righted herself, and, despite being heavy with sea water, caught enough gust in her mainsail to blow her forward.

Still shaken by their brush with death, the men were all for taking down the main mast as well, but Frobisher forbade it, calling on them to ride before the storm on the strength of the sail. The *Gabriel* stayed on course and rode out the blow. The hold was pumped of water, and a new mizzen-mast fashioned from stores and erected on the deck, allowing the ship and her hapless crew to forge ever deeper into the world of ice.

On July 29, Frobisher and the crew caught sight of a headland in the distance, the tip of what we now know as Baffin Island. Spyglass in hand, the former pirate swung his eyes to the north. A similar promontory on a parallel shore came into focus. He had found the channel through the Americas, the passage to the East. He had found it! His eye swung back to the southern, closer point of land.

"I shall call this Elizabeth Foreland," he said, "for surely this is the Northwest way where my Queen's majesty will rule supreme." Nearing its entrance, however, the *Gabriel* was forced to turn back from the channel, so thick was the waterway with drifting pack ice. Instead, the ship tacked restlessly back and forth, always keeping in sight of the coast, convinced that they were looking at America to the south and Asia to the north. They could only wait for the weak July sun to melt the ice or the shifting winds to chase it from the channel. For sixteen days Frobisher demanded that they sail between the two points of land while the season waned and the grumbling of the men grew louder. The *Gabriel* was now a vessel of discontented sailors ruled by a megalomaniac.

It was not until August 19, 1576, that the channel was clear enough to enter and the ship finally ventured into Frobisher Bay, the 240-kilometre-deep fjord that bisects the lower portion of the largest island in Canada. Even as they sailed deeper into the channel and noticed the waterway narrowing, Frobisher still believed he had found the Northwest Passage. Eventually, he decided they should put ashore and hike to the highest point of land to estimate how much water lay between them and China.

Frobisher and six men ventured up through the loose gravel slopes, wobbly and weak-kneed from months aboard a pitching vessel. At the crest of a hill on the northern shore, one of the men came to an abrupt halt.

"Look!" He pointed. "There."

Below them, coming from the opposite shore, were seven boats, each paddled by a single man.

"Make haste," commanded the captain as the men made a rapid descent to their ship. By the time they reached the vessel, the strangers had pulled alongside. With one hand on his pistol, Frobisher approached the strange men with caution. Although he was shocked by their appearance, they seemed to be making gestures of friendship. Each had a broad face and long, black hair. One or two had markings on their cheeks and across their noses, and all were dressed in the skins of hairy animals. Their teeth, though small and sharp, were not bared, and the sounds coming from their mouths seemed nothing like the noise that might be made by a trapped or angry creature.

Frobisher's memories of sailing in South America returned to him, and with a slow, fluid motion, he drew from his pocket a bright silver bell and gently shook it. The sound in the stillness echoed on the Arctic air with a brilliant clarity. A murmur went up from the tawny strangers, a chorus of approval, and Frobisher knew that he and his men would be unharmed.

Trading began immediately, and continued most of the afternoon. Trinkets of polished wood and silver, and even old brass nails from the British vessel, were exchanged for luxurious furs from the men in the skin boats. While these activities seemed to establish some levity among his crew and a general lifting of spirits, Frobisher was quick to caution his men not to trust the strangers.

An exchange of persons was made. The sailors aboard drew lots and the one with the shortest straw disembarked and was paddled to the farther shore where the aboriginals had their camp, while one of the strange company was reluctantly brought on board the *Gabriel,* where he was given salted beef and red wine. When he tasted the food, his eyes rolled back in his head and he puckered up his face, evidently not enjoying the captain's hospitality. The confines of the vessel also seemed to aggravate the man, and he began making motions as if he could not breathe. Frobisher gave the Native a spoon and a small piece of copper piping as he departed the vessel.

"What are the country people like?" he demanded of the shaken seaman who had returned from the settlement.

"Horrible, Sir." The man shook his head. "Too horrible for words. They eat their flesh uncooked and the blood runs down their chins with much laughter. I thought they would be eating me, too, but with no words to beg mercy, all I could do in their dark tent was pray for God to deliver me."

"Was there lots of meat?" asked one of the crew.

"Aye, and lots of men and women gathered to feast upon it."

"Uncooked?"

"Aye."

"Yet with laughter?"

"And much merriment," responded the sailor, eyes wide.

Frobisher's eyes narrowed. "We'll anchor to the south and prepare in case of an attack. My guess is these country people would like nothing better than to take our ship and sail to Christendom. They'd leave us bones upon the beach in an instant if they got the chance."

Frobisher was right to be wary of the Inuit, but he was also quick to condemn them, regarding them as little better than beasts. He would likely have had nothing to do with the country people, as he insisted on calling them, if he hadn't needed their knowledge of the land. More than trade, Martin Frobisher wanted information.

The Inuit were quick to offer seal pelts, and even the pelt of a white hare, but where were their maps of this water and the land that surrounded it? Where were their charts? Frobisher needed to know. These people looked Oriental, with their almond eyes and dark skin. Surely the Orient was at hand. In his cabin, after eating some of the last of the salted meat, Frobisher decided that he would have to go ashore if he wanted any of his questions answered. He needed to know how much longer it would take to reach the Western Sea.

The following day, August 21, 1576, the strangers again approached the *Gabriel*, but this time they arrived in an umiak, a large open boat. One man, older and smaller than the rest, was given as a hostage, and, with great care, Frobisher made his way to the mainland. He was led by the hand to a village composed of a number of rough houses made of stones and skins. No timbers existed on the barrens. Entering the Inuit dwellings, he, too, was invited to feast, but politely declined the offerings of seal fat and what appeared to be raw fish and animal entrails.

In clumsy sign language, Frobisher appealed to a man who seemed to be the leader, asking him to navigate the ship through the waters to the land beyond. Whether the Inuit understood the arrangement is impossible to say, but he appeared alongside the *Gabriel* later in the afternoon and was taken below to the captain's quarters where he amused himself with the bright tools and spy glasses, smiling at his own distorted reflection in the brass instruments.

For a......

more beautiful world.

"Tomorrow we will make for Cathay and the East Indies," said Frobisher to his men as the early darkness descended. Weakened by scurvy and discontented with Frobisher's intention to continue northwest, the crew would hardly have greeted the news with joy.

Confident that the Inuit man would assist them, but not wholly trusting a heathen to sleep among his crew, Frobisher bade five sailors set him down on the rocks of a small island astern of the *Gabriel* but within the master's vision. There was a jostling of men to the launch. Taking the Inuit among them, the six rowed past the appointed rock and beyond. It was the last Frobisher saw of his five crew members.

Historians are divided on the intent of Frobisher's men. Were they abandoning their obsessed leader to take their chances among the people of Baffin Island, or were they captured by the Inuit and held by force out of sight of the ship? The condition of the men points to the theory of desertion. Frobisher waited three days, sailing his ship as close to shore as he dared. Despite sounding trumpets and firing guns, none of his men came forward to join him. They could not, or would not, be recovered, and Frobisher felt deeply betrayed by the brown-skinned men who at first had seemed so friendly.

He sailed on for three days but, reduced to a crew of thirteen, many of whom were no longer capable of hard work, he soon realised that he would have to return to England. With a heavy heart and a deep bitterness against the savages who had killed his dream by stealing his men and his launch, Frobisher sailed back up the strait. The *Gabriel* passed the spot where the men had disappeared. The camp was gone. Nothing was left. It was as if the country people had never been there. Nothing but the wind, tinged with the coming winter, blew across the site that had so recently housed a community.

"They are lost," he confided to his crew, "and we, too, must hasten home or we shall be lost also."

It was a hard admission from a man who had just discovered what he believed to be the Northwest Passage, but Frobisher was not one to skulk back to England. As he journeyed up the strait he saw a group of Natives, numbering some twenty or more, coming toward the *Gabriel* in two umiaks and a number of kayaks. Dropping sail and slowing the vessel, he allowed them to come close, but not before hiding a small cannon at mid-ship, just below the rail. He intended to shoot a hole in one of the great boats and take some captives as ransom for his own men.

Frobisher told his crew to stand down, then greeted the Natives like old friends. To them he must have appeared to be alone, unarmed, and unafraid. As one kayak approached, Frobisher recognized the face of the man who had attempted to drink the wine aboard the ship. The man made a gesture of friendship as he came alongside, but Frobisher only continued to smile. He would not be fooled again by this deceptive race of heathens.

He called his mate to bring a bell, the same one that had

captivated the strangers when they had first made contact. He held it out momentarily, dangling it in the air just beyond the man's grasp. As the Inuit man reached for the bell, Frobisher wrapped his hand around the man's wrists, and, pulling both him and his boat clear of the water, hauled them together over the rails and onto the deck. It was an astonishing display of strength, and a great shout went up from the assembled boats. Many of the paddlers made for shore quickly. As they went, they shouted and howled in their strange tongue that one of their own had been taken.

On board was chaos. The Europeans surrounded the Native man, who crouched in fear. Frobisher held his fingers aloft by way of translation. "I will give you back in exchange for my five," he said to the cowering man. "Five for one." But the man's fear overwhelmed his understanding, and he only crouched lower. Frobisher locked him in a prison below decks, and waited at anchor, hoping for word of his men. None came.

On August 24 snow fell in the evening, and Frobisher could wait no longer. At least he had a human specimen to bring back to England, and he bade each man in his company venture to the shore to bring back some souvenir of the passage to Asia. The men wanted only to get under way, but they each reluctantly scooped up a few black rocks, which rattled around in the hold all the way back to England. Forty-four days later, the *Gabriel* arrived at the entrance of the Thames.

Frobisher, who had been reported lost by the master of the *Michael,* became the toast of London on his return. He had been gone for four months. He had returned with news of the Northwest Passage. He had also returned with a pale and emaciated savage from the hills of Asia, who quickly succumbed to the unfamiliar, germ-laden air of central London, and a boat the likes of which had never been seen before. The thing that really caught the attention of the Queen and her advisors, however, were those few black rocks the sailors had picked up, for they were rumoured to be laced with gold.

Michael Lok was dubious about the stones. Lok's assay-master found them to be iron pyrites, not gold at all. A trace of the precious metal *may* be present was all he would concede. Another metallurgist came up with the same discouraging news, but Frobisher was sure he had discovered gold. In any case, the chances of backing for a second voyage would be certain only if an alchemist—or a rumour monger—could change what others had diagnosed as marcasite into pure gold.

Frobisher found his man in the person of one Giovanni Agnello, an Italian assay officer. Despite the informed opinions of others, Agnello confidently pronounced the metal in the black rocks of Baffin Island "to contain much gold." In less time than it took for the poor Inuit captive to fall sick and die, gold fever had gripped the nation.

So strong was the pull of gold on the English imagination that Frobisher himself appears to have succumbed to it. But gold or no gold, he was now assured of a return voyage to the land of ice that opened to the East. He was also assured his place in history. He saw himself lauded and esteemed in the Elizabethan courts, and fêted throughout the world as the man who discovered the Northwest Passage.

The winter of 1577 dragged on. Brash and boisterous in public, Martin Frobisher had plenty of galas and gatherings to demand his attention, but the rogue and braggart had private concerns about the next voyage. His plan to get back to the Arctic in order to sail up the channel he had discovered had backfired. Gold was the word on everyone's lips now. Gold from the New World. Gold from the Arctic. Gold for the Queen. Finding a way to the East had taken a distant second to the prospect of bringing back gold to England. His second voyage would be a mining and hauling operation, not an expedition of discovery. Even so, Frobisher was still keen to get back on the water. He was riding on his fame, and had recently been commissioned High Admiral of the Waterways of Cathay.

The Queen, impressed by Frobisher's fearless determination, was donating capital of her own to the adventure, as well as a vessel. The *Ayde,* ten times larger than the *Gabriel,* was given toward the expedition, and the two ships from the previous year refitted. On May 25, 1577, with Frobisher aboard the *Ayde* and trusted masters piloting the other two vessels, the entourage of 125 men set out on the same west by northwest course.

Frobisher's sailing orders were to fill the largest vessel with gold ore and to send one of the smaller ships up the passage to a point about eight hundred kilometres from the entrance, where a party would over-winter. As the *Ayde* returned to England, heavy with treasure, the explorers would make their way to the Orient on the two smaller ships when weather and ice permitted.

Just thirty-nine days off the British coast, the lead ship, the *Michael,* fired a cannon to indicate that land had been sighted. And there they were again, the great ice cliffs of Greenland. Frobisher felt his heart race. The voyage to this point had been relatively uneventful, and the company was in good spirits as they rounded the tip of the continent of ice and made their way toward Frobisher's strait.

The weather turned, of course. Fog descended, to be interrupted only by blasts of boreal wind and snow. The *Michael* lost her topmast in high seas, and the sailors hugged their cloaks to their chests and blew on their fingertips to restore sensation. This cold, mid-summer weather was much like what they experienced in the darkest months of winter at home.

After four days of creeping through fog in a navigational guessing game, the wind picked up and cleared the air. Frobisher was shaken when he realized how easily one of the ships could have been crushed or snapped in two had it struck any of the monstrous icebergs that were drifting southward against their course.

Returning to the island where he had forced his men to collect souvenirs, Frobisher could find no place to land the mighty *Ayde*. It drew too much water to slip into the previous harbour. A reconnaissance trip in one of the smaller vessels found a neighbouring island rich in the black rocks, but without an ice-free harbour to anchor the main ship. Miners were dispatched to the shore to pluck the ore from the land while Captain Frobisher and a few of his officers walked to a hill some three kilometres inland. At the crest of the hill, Frobisher halted.

"We'll build here a cairn with a cross atop and call this place Mount Warwick," said Frobisher slowly. "And we'll pray to our Father in Heaven that we subdue the beastly people of this land and bring them to knowledge of our Saviour Christ Jesus."

The men, bent double in the wind that seemed to blow straight through them, spared a quick glance at the landscape Frobisher had indicated. Emptiness was all that met their eyes, the desolation of the barrens, not a tree or a stick. Who could live in this brutal clime? Nonetheless, they built their cairn. A trumpet was sounded once to mark the ceremony, and the men clambered down the bluff toward the sea. As they were approaching level ground, a man appeared, gesturing wildly and shouting in a guttural tongue.

Frobisher answered with a second trumpet blast, and the stranger leapt and seemed to dance with a nimble gaiety. Frobisher scowled, for he was convinced the Natives were all possessed by demons. One moment they would be laughing, and in the next they would lay slaughter to their fellow humans, butchering them for meat.

Raising two fingers to the Inuit, Frobisher indicated that two of his men should go forward to meet two of the Inuit, for a group had appeared at the sound of the horn. The four met in sight of all, and an exchange of goods—a bow from the Natives and a few straight pins from the midshipmen—seemed to satisfy the curiosity of both groups, but neither cared to venture further with the other, despite invitations issued on both sides through sign language.

"There is no trust in them," cautioned Frobisher, when his party returned to the area where their launch was pulled up. "We should take one or two of them as prisoners. These men know the waterways and could lead us on to China. Look, two approach now."

The Inuit men approached Frobisher and his lieutenant, George Beste, keeping their eyes averted. Beste and Frobisher were also careful to avoid eye contact, and kept their heads lowered until the two strangers were abreast of them. Then, as though at a prearranged signal, the Englishmen lunged. Grabbing the men around the waist and shoulders, they tried to pull them into the launch. But Frobisher slipped on the ice and his victim scrambled away. Beste was unable to get a hold on his opponent's slippery sealskin coat, and the man soon twisted away and ran for the rocks. There the two Inuit men armed themselves with bows and arrows and charged back. Frobisher and Beste clambered into their boat and took cover in the gunwales. An arrow found its mark in Frobisher's upper thigh.

"I've been hit!" he cried, as the pain surged through his

buttocks and down his leg. "Those heathens have struck me with their weapons!"

A shot from the gun mounted on one of the boats soon sent the Inuit scrambling for the hills, but not before one of the landing party, a Cornishman named Nicolas Conyer, gave chase. He grabbed one of the Inuit and wrestled him to the ground. A quick blow to the head subdued him, and Conyer slung his trophy across his back and shoulders and carried him to the waiting vessel.

"This is what ye wanted, Sir? A man to come aboard and help us navigate these icy shores?" asked the burly seaman. "This one should do well enough, Sir. The other, the bigger one, got away, but this fellow should be able to assist us if we need it." Conyer unceremoniously threw the Native into the back of the dory, where he lay unmoving, either unconscious or paralyzed with fear.

By the time Frobisher and his men were ready to make for the *Ayde*, the wind had shifted and they could make no headway in the rough seas. They spent an uncomfortable night on a nearby island, with only a small fire to challenge the sense of foreboding that flesh-eating humans inhabited the land around them.

Regaining the ship the next morning, Frobisher decided it would be better to head farther up the strait, to a place where the iron ore was more accessible. Once the anchors were set—in the place named Jackman Sound—Frobisher left the *Ayde* and took one of the smaller ships, the *Michael*, further up the bay to continue his exploration.

On July 29, almost 120 kilometres from the mouth of the bay, he discovered an island, an excellent site not only for mining but for anchoring the ships. He named it the Countess of Warwick Island and immediately despatched the *Michael*'s miners with picks and shovels. Veins of gold were apparent even to the naked eye in the mineral-laden ore, and Frobisher knew he had struck it rich. He quickly sent for the *Ayde* in order to bring back as much of the gold as possible.

While the *Michael*, piloted by Captain Yorke, sailed back for the *Ayde* and her crew, Frobisher stayed on shore to supervise the stone-gathering operation. Lieutenant Beste and a smaller company ventured inland to learn what they could about the place where they had landed. The Inuit captive, a rope around his waist, was taken along as a guide. Presently, the party came upon an abandoned Inuit camp. The prisoner gave a sob of recognition and sank to the ground. He refused to move, apparently overcome by his memories.

The dwellings, made of skins stretched over whale bones, mystified the British, and only two dared enter them. "They are round like ovens, having holes dug out at the front so that men might enter like foxes," reported one of the sailors. "Each house has only one room, with half the room dug into the earth and the other half raised a foot higher with broad stones. It seems it is here, up on the shelf, that they make their foul nests, and down on the ground they do their beastly feedings, for there are bones lying about in these defiled dens. I longed for sweet air." He filled his lungs in the crisp autumn breeze and continued: "'Tis no wonder they have gone, their caves being too foul for breath. Look." He pointed at the small Inuit man. "Look at what he is doing."

The company turned to see the captive squatting over a design he had made on the earth: five small sticks were placed in a perfect circle, and at the centre was a bone.

"It's witchcraft," said one of the men.

"Aye, he's calling on the Devil," said another.

"Nay, he's leaving a mark for his people to see he's still alive," ventured a third.

Beste looked closely at the design. "See how the five sticks surround the bone? Likely, he wants people to know he is a captive, and the five outside sticks must represent the men that were lost and captive at this time last year." Beste suddenly grew excited. "We must tell Captain Frobisher of this, for, if this means what I believe it does, our men may still be alive."

The party made its way back to the mining area to discover the *Ayde* now in the harbour. A conference was convened, and news from Captain Yorke added to the possibility that Frobisher's five men had somehow survived the winter.

"I was on my way here," he reported, "when we anchored the last night in a small bay. There still being twilight, and I, discerning shapes from the bow, did put ashore to find a camp of the country people most recently deserted."

"What did you find at this camp?" Frobisher demanded.

"We saw newly killed flesh of an unknown sort and the bones of dogs and carcasses of strange animals," Yorke replied. "But it was in the tents we saw the most marvellous thing."

"Yes, yes. Quickly now."

"There was a doublet of canvas made after the English fashion, and a shirt, a girdle, and three shoes for contrary feet of unequal bigness; apparel, I should think, of our five comrades who were intercepted last year by these savages."

Frobisher quickly decided that Yorke's discovery warranted further investigation. Leaving Lieutenant Beste in charge of the mining operation, Frobisher set forth for the place Yorke had mentioned with some thirty-five men in two rowing pinnaces.

On landing at the deserted camp, the British party split in two, with half heading inland to scout for signs of activity, the other half staying with the boats. In a valley not far from the landing area, a new camp was discovered with some sixteen to eighteen persons and four small tents. Frobisher decided the Inuit should be captured and brought back to the mining site. The Natives tried to flee to their boats, but the sound of gunfire alerted the mates of the *Ayde,* and more men soon appeared. The Inuit turned on Frobisher's men, casting arrows and darts in a desperate attempt to defend themselves. But they were no match for British firearms. Six Inuit were slain. Some threw themselves headlong into the sea to avoid capture. The rest escaped among the rocks. A young woman with a child on her back was found clinging to a rock as if trying to dissolve her body into the stone itself. Assuming she was a man, a soldier shot at her, breaking the child's forearm. A shriek of pain issued from the baby's mouth, and Frobisher's men quickly took the woman and the child prisoner.

"We will give her as a comfort to the first savage," said Frobisher after directing the ship's surgeon to apply ointment to the child's arm. "There will be company in two, and we can teach this child the English tongue."

They dragged the woman carrying the child on board the *Ayde* and thrust her into the same cabin where the Inuit man was locked up. A few men watched from the doorway, speculating on how the two might engage. Lewd comments were heard, but they gradually diminished as the sailors witnessed an extraordinary communication between the man, the woman, and the child.

For the first two days there was no exchange of language between the two adults. It was as if their words had been stolen by the shock of being captive in a strange place surrounded by leering white faces. Only the baby made any noise, and it sounded like a kitten, forlorn and pitiful. At length, the Inuit man took the wounded child from its mother's arms and held it against his heart. The child slept while the mother gazed into space.

On the third day, the woman began to sing in her strange language, and the man responded with a long story to which the woman listened, occasionally nodding solemnly. They made no eye contact. The woman sat with the babe in the crook of her arm, silent and still except for the occasional nod of her head or the appearance of a small, pink tongue, to lick at her child's wound. The sorrow in the room was palpable.

On Monday, August 6, with the mining operation in full swing and the area properly fortified, a contingent of Inuit appeared at the summit of a nearby rock face. They seemed to want the mother and child back. Frobisher brought her out and placed her in full view of her people, securing her ankle to a rock. He then moved forward with the Inuit man and proceeded to negotiate an exchange: he would give back the man, the woman, and the child for the five men who had disappeared the autumn before. Pointing to the sun and indicating a three-day journey, the Natives requested that Frobisher send a letter. He wrote it at once:

> *In the name of God, in whom we all believe, who, I trust, has preserved your body and souls amongst these infidels, I commend me unto you. I will be glad to seek by all means you can devise, for your deliverance, either with force or with any commodities within my ships, which I will not spare for your sakes, or anything else I can do for you, I have captive of theirs a man, a woman and a child, which I am contented to deliver for you: but the man I carried away from hence last year is dead in England. Moreover, you may declare unto them that if they deliver you not, I will not leave a man alive in their country. And thus unto God, whom I trust you do serve, in haste I leave you, and to Him we will daily pray*

for you. This Tuesday morning, the seventh of August, anno 1577,
Yours to the uttermost of my power.
Martin Frobisher

I have sent you by these bearers pen, ink and paper, to write back unto me again, if personally you cannot come to certify me of your estate.

The Inuit reappeared at the crest of the hill three days later, as promised, but there was no sign of the lost Englishmen. Frobisher's men noted that only some of the Inuit showed themselves; the rest hid among the rocks, as if hoping to mount a surprise attack. They had no news, it seemed, and were using delaying tactics to keep Frobisher's interest. Frobisher decided that his men would surely have sent word if they were able to; in all likelihood, they had been consumed by the flesh-eating people. There seemed little point in carrying on any further dialogue with the Natives.

Frobisher returned with his men to the mining area, where some two hundred tons of ore had been loaded aboard the *Ayde*. With ice crystals in the harbour and ice forming around the bellies of the ships at night, he realized it was time to head back to England. Laden with gold for his beloved Queen, but yet with little new knowledge of the passage to Cathay, the three vessels set sail for home.

They reached England amid much fanfare in the middle of October 1577, laden with ore that Frobisher was convinced heralded a new and splendid future for Britain. With his lost men a distant memory and the Inuit family to put on display, he was the triumphant man of the hour. A third expedition was already being planned, and this time Frobisher felt certain he would sail through to the East.

The Queen decreed that the new land be called *Meta Incognita* and, as her refiners and assay officer lined up to determine the amounts of gold in the northern rocks, she bade Frobisher prepare a bigger and better expedition to bring back information on the land that lay beyond the horizon of human knowledge.

In May 1578, Martin Frobisher embarked on his third and final voyage to *Meta Incognita*. By all accounts, the passage was horrific. Rounding the southern tip of Greenland, the ships sailed into a sea of ice. One of the larger ships, the one-hundred-ton *Dennys*, struck an iceberg and sank just moments before her crew was pulled aboard the flagship, the *Ayde*. Days later, a storm from the southeast arose with sudden fury and blew the pack ice against the boats. The boats, in turn, were blown toward land, where they were forced up against the shore ice. The sailors hung beds, planks of wood, chunks of cable, whatever they could lay their hands on, over the sides of their ships to deflect the grinding pieces of ice that threatened to slice the vessels in two. Many hands stayed on deck all night through the storm, pushing back the ice floes with rods and oars until their hands bled and they sank to the deck with exhaustion.

The storm broke up the fleet, and Frobisher had no idea

how many ships had survived it. The majority of vessels followed the *Ayde,* with Frobisher aboard, 290 kilometres up Hudson Strait before the captain abruptly declared it a navigational error and dubbed the waterway Mistaken Strait. Thus ended the most successful portion of the voyage, in terms of exploration. The ships hove to just as ice fog descended. In a near white-out, the fleet threatened on all sides by ice and shore and collision, Frobisher led his entourage on a circuitous and dangerous course back to his strait and the iron ore that was so coveted by the Crown.

Just when the crew thought they could begin to relax and proceed with gathering ore and colonizing the area, a huge island of ice struck the *Ayde*'s bow and drove the anchor fluke through the hull. Water poured in below decks. The ship began to list. Had it not been for a constantly manned pump and the quick action of the ship's carpenter, the *Ayde,* already carrying a double load of hands, would have met the fate of the luckless *Dennys.*

At last, on July 30, with six inches of snow on the hatches and the sailors weather-worn and weary, the fleet assembled at Countess of Warwick Island and began the mining operation. Once again, plans for a group of men to over-winter were thwarted. Much of the building materials, with which they had planned to erect a fort, had been lost when the first ship sank. In any case, the summer weather was such that men could scarcely imagine the winter, and the one-hundred-person contingent that had agreed to colonize the barren lands quickly changed their collective mind.

The ships were stripped of all but the necessary provisions to get them home. Lumber and local stone were used to construct a house on the island. It had two purposes: first, it would tell them whether a European-style structure could withstand the forces of winter in *Meta Incognita*, and second, it was to act as a positive talisman for the infidels. Frobisher had the house filled with trade goods—small mirrors, bells, dolls, glass beads, nails, all manner of pins and iron scraps, and bright bits of fabric. An oven was also built inside the structure, and Frobisher ordered bread baked just prior to the fleet's departure. No doubt he imagined the Natives, lured by the smell of baking bread, would become more favourably disposed to the white men when they returned for good.

At the end of August, nine of fourteen ships sailed for England loaded with ore. By the time the remaining five were patched and their holds loaded, some thirteen thousand tons of rock were bound for the British Isles.

It was a blunder of great magnitude. As he shepherded his fleet into the harbours of his homeland, Martin Frobisher remained unaware that the ore he carried was as worthless as his navigational skills. As Michael Lok's assayer had concluded after Frobisher's first voyage, it was not gold at all, but iron pyrites—fool's gold—an opinion that was confirmed shortly after the third expedition set sail. Frobisher's jig was up, his reputation in tatters.

The ore, mined at such cost, was unceremoniously dumped into the harbour. Some of the ships used in the explorations, including the veteran *Gabriel*, were sold to pay off creditors, and the mariners, who had suffered much on the third voyage and promised money aplenty when they got to shore, remained unpaid.

Refusing to be crushed, Frobisher returned to what he knew best: piracy, and the sea life of the fighting sailor. He did a brief stint with the British Navy, acting as Sir Francis Drake's second-in-command in the 1585 West Indies Raid. But second-in-command was not Martin Frobisher's style, nor were the tropics his choice of adventure. He longed for another voyage north, to find the passage once and for all. To that end, and despite his reputation, he even managed to raise some capital. But the dream finally died when the command of such an expedition passed out of his hands.

He had found a strait—which proved, in fact, to be a deep-water bay—on the coast of Baffin Island. He had found, but failed to probe, the Mistaken Strait, which would soon be known as Hudson Strait—the key channel to Hudson Bay and the undoing of many subsequent Arctic explorers. He had mined the aptly named fool's gold. He had made blundering, foolhardy, vain, and pig-headed decisions. He had antagonized, brutalized, and grossly misjudged the people of the Arctic. Despite this record, Martin Frobisher was still known in the sixteenth century as an authority on the northern waters. He was the man who had ventured furthest into the icy mysteries at the top of the globe. This, above all, is Martin Frobisher's legacy.

HENRY HUDSON

The Mutiny

Henry Hudson was martyred for the cause of the Northwest Passage, cast adrift by his own men and left to die on the shores of Hudson Bay while his ship, the *Discovery*, navigated through the pack ice and made for England. Abandoned with his young son, the ship's carpenter, and a handful of sick and dying seamen in 1611, Hudson will forever be remembered as the man who penetrated the great inland sea of Hudson Bay and over-wintered there, only to be left to die by mutineers.

Hudson was not the first explorer to venture north since Frobisher. A number of voyages had been undertaken by the English mariner John Davis from 1585 to 1587, and again by George Weymouth in 1602. The dream would lie dormant for years, it seems, then suddenly flare up again for no apparent reason. Each successive voyage garnered a little more information, however, and each added its observations, calculations, and documentation to the polar map that was steadily becoming more detailed and more accurate.

John Davis, for instance, with neither spectacle nor splendour, gave the world a more complete map of the northern latitudes, including the great waterway between Baffin Island and the eastern coast of Greenland, subsequently named Davis Strait. He discovered the large inlet above Frobisher Bay, Cumberland Sound, and, deeming it an unlikely thoroughfare to the East, sailed south to make his most important discovery, proving that Frobisher's

Mistaken Strait was actually the entrance to Hudson Strait, which was, in turn, the entrance to the great bay.

Davis had noted the Atlantic tides rushing into the gulf and what appeared to be an ice-free stream of water surging into the broad passage. Convinced he had found the Northwest Passage, he determined to continue exploration the following spring. He would have mounted another expedition, had it not been for the Spanish Armada. With England at war with Spain, all Arctic exploration was abandoned. Davis could only pass along his findings in detailed charts and two books about his journey, *Seaman's Secrets,* published in 1594, and *Hydrographical Treatise,* published the following year. Both these works supported the notion that the Northwest Passage lay between latitude 62 and 63 degrees, the opening to Hudson Bay.

In 1602 George Weymouth began the trip through Hudson Strait in a fifty-five ton vessel, the *Discovery,* but ice conditions forced him to turn back before he reached the bay. His observations were simple—"No land barrier lay ahead"—and he eventually donated his sturdy ship to the sailor who was to give his own name to the strait and the bay, Henry Hudson.

Hudson was not unknown to financiers and promoters in the British Isles. A professional sailor, Hudson was not only convinced of the existence of the Northwest Passage, but that *he* was the person who would first sail it.

In 1607, five years after Weymouth had returned from his unsuccessful attempts to penetrate the "Entrance to the

East," Hudson persuaded an influential group of businessmen, the Muscovy Company, to sponsor a trip across the top of the world. This group, by the turn of the century, had gained a monopoly on trade within the White Sea and the Siberian coast, and listened with interest to Hudson's proposal. He would sail to China, he claimed, by heading north to the pole. The central polar seas, believed at that time to be ice free, would open up to the East, and the Muscovy Company, through him, would secure dominion over that trading route as well.

Hudson was a good sailor, but a less-than-persuasive negotiator. The company agreed to sponsor his expedition, but certainly not in style. He was given an old hulk of a ship, the *Hopewell,* that had been part of the Frobisher fleet some twenty-eight years earlier, with a crew of ten and one cabin boy. They sailed north to a latitude of 80 degrees, the northernmost waters ever entered by a British vessel. Between the islands of Spitzbergen and the east coast of Greenland, however, the *Hopewell* confronted an unbroken wall of ice. Hudson could go no farther.

He turned his ship around and nudged the vessel between the islands of Spitzbergen, which had been discovered by the Dutch explorer Barents in 1596. While there, he noted the abundance of whales, walrus, and seals, and brought back news of a rich fishery to his sponsors. The company was delighted. Hudson had not discovered the passage across the pole, but the whaling profits from Spitzbergen were more than enough to have made the venture worthwhile. With their attention diverted from exploration to harvesting and processing the fishing grounds of the northeastern Arctic sea, Hudson could find his own way to the East, as far as the Muscovy Company was concerned. Captain Hudson, anxious to get back out on the water, would have to find another organization to underwrite his voyage of discovery.

The Muscovy Company, however, must have felt some allegiance to the man, for they did not leave him totally high and dry. The company sent him on one final voyage, and in 1608 Hudson attempted to find the passage by venturing northeast across the top of Russia. There, while trying to force the *Hopewell* into the ice-clogged Kara Sea, he was met with threats of mutiny from his crew. The leader of the insurrection was one Robert Juet, a man who, less than two years later, would prove to be Hudson's undoing.

Seeking new sponsorship but finding none in Britain, Hudson approached the Dutch East India Company, which was known to be favourable toward Northwest Passage exploration.

"There has been a new channel located by the Italian, Verrazano," Hudson told the merchant company. "I believe it is the passage through to China and I beg you, give me the opportunity to explore this inland waterway."

The Dutch agreed to finance a voyage, and again the *Hopewell* was put into service, sailing southwest across the Atlantic and up the Hudson River. As the waters became shallow and it became increasingly apparent that the

Northwest Passage was yet to be discovered, he sailed back to England, but not before landing a colony of Dutch settlers on the banks of the Hudson. New Amsterdam was the first name given to the settlement that later gave birth to New York City.

Back in England, Hudson's reception was chilly. He had sold his services out of country, abandoning the British cause to forward that of Dutch exploration. The English were appalled. It was, in their minds, a kind of national betrayal. Were it not for a consortium of scientific-minded and enterprising men—Sir Dudley Digges, Sir Thomas Smith, and Mr. John Wolstenholme—Henry Hudson would have never sailed again under the British flag. These men, however, were intrigued by the venture up the Hudson River, and believed that a passageway across the continent of America was certain to be found soon. Henry Hudson was, in their minds, the man to lead the search. Thus, in April 1609, after forcing him to take a vow prohibiting him from hiring his services out to foreigners, they employed him to sail north by northwest on the *Discovery*.

Now Hudson needed a crew.

Little has been written in history books about Hudson's character, but it seems fair to say that he was not an astute man when it came to the discernment of personalities. He was likeable, and felt none could dislike him. He was naive in many ways, and strangely guileless. But like his predecessor, Frobisher, Henry Hudson was obsessed with finding the Northwest Passage. The search consumed him, and superseded sound judgement.

The man Hudson chose as his first mate was the same man who had threatened mutiny not two years before when the *Hopewell* was seeking a passage through the Siberian waters. Robert Juet presented himself, claiming to be as determined as Hudson to find the gateway to the East.

"I would be honoured to sail with you again," Juet confided in a dockside tavern in London. "You know yourself, I've tasted the northern waters and I'm not afraid of the ice and cold."

"True," conceded Hudson. "And you're handy with navigational instruments?"

"The best in these parts," said Juet. "I hear you've got a sound vessel."

"I've been given the *Discovery*. She's fifty-five tons, and her hull is planked in pure oak. She's small, but has a shallow draft. She'll get us through the ice."

"And who's to navigate?" asked Juet, delighted at how quickly he'd been drawn into the adventure.

"I've a man named Robert Bylot. He's a quiet sort, but well versed in matters nautical." Hudson leaned forward conspiratorially. "We'll find it this time, Juet. The passage is to the northwest. I've this to show you."

From beneath his cloak Hudson extracted a dog-eared document. It was John Davis's *Hydrographical Treatise,* written more than a decade ago.

Juet's eyes narrowed suspiciously, and he reached for the book. "What's that?" he demanded. But Hudson withdrew the book. He may have had a sudden memory of Juet's hands reaching for the wheel of the *Hopewell* the previous year, turning her around, and announcing that he and the crew would go no farther. But Henry Hudson was a man who believed in the good of his fellow human beings, and as he tucked the book under his coat, he could not help but reflect that the last voyage *had* been a futile adventure. There was no Northwest Passage in the northeastern waters. Juet had been right to turn them around. The passage lay north by northwest and, by God, this time, with a solid vessel, the sponsorship of moneyed merchants, and the information in Davis's document, the expedition was sure to succeed.

Hudson rose. "We sail in four days, Juet," he said. "Good-day, Sir." And with that, Henry Hudson re-hired the man who had so recently betrayed him.

Robert Juet was not the only suspect character on Hudson's final adventure. As the ship sailed down the Thames to open sea, its final port of call was the city of Gravesend. There, against the authority of Hudson's sponsors, another man was taken aboard as the ship weighed anchor.

There are a number of conflicting accounts of how Henry Greene came aboard the *Discovery*, but most agree he was a bad piece of business. A young man beautiful in countenance but vile in action and personality, Greene apparently stole or was smuggled aboard the *Discovery* after a fight with his father. The handsome young man quickly ingratiated himself with the captain.

Greene soon became an intimate part of the twenty-one

person crew—perhaps too intimate; accounts tell of his "going into the fields" with different men—which included Hudson's son John and his sponsors' representative, a defrocked priest by the name of Abacuk Pricket.

By the time the *Discovery* had reached the Shetland Islands on April 27, 1610, the diverse personalities aboard ship had begun to clash. A storm that blew in from the northeast and forced the small vessel to take shelter on the lee side of Iceland was enough to make tempers flare. Greene and the ship's surgeon, one Edward Wilson, had a vicious argument, which resulted in a fist fight that divided the crew. Juet called Greene a spy, whereas Pricket defended the lad, saying that Wilson had a tongue that would "wrong the best friend he had."

Hudson sided with Greene and threatened to send Wilson back to England on the next whaling vessel they met. The fact that the captain had chosen to defend the young stowaway instead of the respectable seaman further alienated the crew, and it was with a deepening sense of division and discord that the *Discovery* pressed on to Greenland.

By mid-July they had reached the opening to Hudson Bay, the narrow strait that John Davis had earlier described as "the furious overfall," for the bay was like a huge salt-water lake, its outlet to the Atlantic a massive set of rapids dependent on the tides. To the men aboard the *Discovery*, it looked like a clear, direct way to open water—the Pacific Ocean—once the bare mountains on either side of the strait gave way to the sea.

With Hudson at the wheel, the *Discovery* plunged into the rolling current. They swept past Resolution Island at the south end of Frobisher Bay, and along the rocky coast, all hands engaged in an effort to steer a course that would avoid the pack ice that threatened on all sides.

"Our course is as the ice lies," said Pricket to Juet, as the *Discovery* veered first north, then northwest, and finally to the south and southwest.

"Yes," replied the mate, "and our captain seems not to know or care of our direction. There'll be blood on the decks before this is over, Pricket, and I dare say you'd do well to keep your head down and your sword at your side."

Hudson, meanwhile, was too busy steering a course through the tumultuous strait to worry about the morale of his crew. Robert Bylot, the chief navigator, proved to be irreplaceable at the helm, and his steady hand helped guide the ship to the relative calm of Ungava Bay. Hugging the eastern shore of the bay, the *Discovery* sailed to the cape where the land turned sharply southward and the water stretched as far as the eye could see. There, on August 2, 1610, in a place Hudson named Cape Wolstenholme after his sponsors, he assembled the crew amidships to hear their grievances. With the immediate danger over, the men were calmer, and many of their complaints were now levelled at Robert Juet who had been persistently undermining Hudson's authority.

"We are more than a hundred leagues farther than any Englishman before us has been in this place," said Hudson to his men. "We have met and defeated the ice. From here it is clear sailing. But the time is short. Which men with wrens' hearts would put aside the labours we have spent and the dangers we have conquered to return home when the Northwest Passage is at hand?"

The ship's carpenter, Phillip Staffe, stood quickly by his master's side and uttered the single word Hudson needed to hear. Onward.

The rest of the crew seemed to take heart from Staffe's gesture, and turned back to their positions, hauling on the lines to move the vessel down the eastern shore of the huge uncharted waterway. Juet, deposed as first mate in favour of Bylot, was even less kindly disposed toward Hudson than he had been before, and the grumbling below deck continued.

Anchoring between the mainland and a small island, Hudson despatched six men to a place he named Digges Island to see if they could find game. A number of partridges were discovered nesting in the tall, lush grasses, and a small herd of caribou roamed about the island. The men also spotted an Inuit camp, although it appeared the Natives had recently fled, likely at the approach of the double-masted ship. When the men went to explore the camp, they found a number of gutted and plucked fowls hanging off a line, and they helped themselves.

"We should tarry here awhile, Sir," said John Williams, one of the men sent out. "There's fresh meat to be had, and it seems we'd do well to refresh ourselves in this place."

"We'll carry on," said Hudson, ignoring Williams's suggestion to stock the ship with fresh meat. "We mustn't linger with time so short."

Williams, feeling an ache in his joints and a loosening in his teeth, was frankly worried. These were the sure signs of scurvy. Fresh food and rest was all the remedy he needed, but Hudson was more concerned about making time. He assumed that supplies could be brought on board once they reached the ports of the Orient. Warmer seas, fresh fruit and meat, spices, rum: all these luxuries lay ahead, but to get them they had to press on. Williams reluctantly agreed. The decision eventually killed him.

The *Discovery* was racing against the onset of winter, but it was also racing against a diminishing food supply. The vessel had been stocked with six months' provisions—a ration for each man for 180 days—and they were already three and a half months at sea. Yet Hudson stubbornly sailed south, convinced that he had crossed through the American continent and was heading toward China.

The land bore away to the east, and still the *Discovery* sailed south. For days the ship sailed on, abetted by a constant wind and the heady sense of success. It was not until Hudson reached the southern shore, now known as James Bay, that the water became shallow enough to anchor close in. As it was October 27, and too late in the season to sail back, the crew had no choice but to spend the winter in the bleak, unknown land they had found at the end of the great bay. It was a far cry from the paradise they had expected. They were weary and becoming sick.

Hudson, finally forced to admit that something had gone horribly wrong, instructed the crew on November 1 to turn the ship toward a shoal at the mouth of the Rupert River and prepare the vessel to be frozen into the ice. A reconnaissance trip to shore revealed bird and animal tracks in the snow, plenty of wood for fires . . . and human footsteps along the shore.

Hudson belatedly tried to cheer the crew. "Any man that brings aboard fish, fowl, or game," he said, "will be rewarded with extra provisions. We must do what we can to keep body and soul together, for the winter is upon us and stores are low. Take your muskets and your snares, then, my friends, and hunt for food upon the shores and we will survive this sunless season to sail again."

"'Tis too late for Williams," said the bo'sun, William Wilson. "He was stiff in his bed this morning, grey as the ice mountains yonder and equal in chill."

"What?"

"The gunner," Wilson repeated. "He's dead." He shifted his defiant gaze from the captain to the crew, and raised his voice: "And we might all prepare to meet our God, so poorly laid are the plans for this miserable adventure."

A murmur of agreement came from the crew. This time they would not be stilled. Their situation was all too plain. The weather was closing in. More snow fell each day. The only thing dropping more rapidly than their spirits was the temperature. Now the spectre of death had appeared, and that fearsome apparition would not be banished easily from an immovable ship pinned to the sea by sheets of ice.

Hudson took to his cabin below decks, where he told Henry Greene of the demise of the gunner and the hostile manner of the crew. Greene was not the least sympathetic, but his eyes lit up when he heard of Wilson's death.

"He has a grey cloak of the finest wool," he said, "thick and warm it looked to me, like something I would dearly love to have."

"His possessions will be auctioned at mid-ship in the morning," Hudson replied.

"But the cloak? Surely it could be mine without the bother of the auction."

"What cloak is this? What damned cloak? The man is dead, Greene, and we are fast in the ice."

"I know it all too well. And you, the same. But this grey cloak could keep the chill from my bones, keep the winter at bay, that I might better serve you when the darkness is all around us, as it soon will be."

Hudson relented. It was true, the lad who had boarded the *Discovery* at the eleventh hour was ill-prepared for the Arctic winter. And what was a cloak, after all? It was a small token to pay for the loyalty of one of his favourite men. The gunner would need it no longer. Hudson agreed to secure the cloak for Greene before the auction began. He knew this constituted a breech of trust with the other men and was a direct violation of seafaring tradition. But he felt authority slipping away and was prepared to grasp at any glimmer of loyalty.

The next morning he called for Phillip Staffe. "You're to go ashore," he told his friend, "and build a sturdy cabin for the winter."

Staffe balked at the notion. "But it's November, Sir. The winter is here. This is not the time for building. The ground is frozen, and you yourself know snow is lying thick upon the shore."

"We need a structure, Staffe. You are the ship's carpenter, are you not?"

"Aye."

"And you have tools, I presume."

"Yes, but . . ."

"We *will* have a house built on the shore, and you *will* build it for me."

It was no longer a request, but a command. Still Staffe protested, claiming that exposed flesh would freeze instantly, and how could a man haul lumber and pound nails with his bare hands? In October, yes, perhaps a house could have been built, but by now it would be impossible.

Hudson, stressed beyond endurance, flew into a rage. He cursed Staffe, struck him across the chest, and said he would see him hanged on their return to England.

Fortunately, Staffe, as well as being a large man, was a man of great calm. "I am a ship's carpenter, not a house carpenter," he said, measuring his words carefully. "Yet nor am I a man to disobey my master. You want a house built in the dead of winter, a house you shall have, but I will not let this house become my tomb. The building will be slow to rise in this terrible clime, but, as God is my witness, I will work to keep the peace and restore order in this place lest we all perish."

Hudson's notion to build a shelter on the land may have been the reasoning of a man who knew he would ultimately be abandoned. His authority over his men, always tenuous, was rapidly dissipating, and he knew from experience that mutiny was not uncommon on maritime adventures. On the other hand, he was known for being blind to the needs and notions of the people around him as he focussed instead on his own personal mission. Perhaps his aim in building a house was to establish a symbol of civilization, a monument to his labours, and a permanent record of the success of his voyage and the penetration of the Northwest Passage. Whatever his reasoning, construction started the day after the master's tantrum, proving that Hudson still had some power over his crew.

A few days later, when Greene and Staffe broke from the construction to go hunting, Hudson became outraged. To his mind, the chief builder had abandoned the project. To make matters worse, he had taken Hudson's favourite boy on a fowling expedition. Greene's allegiance was crucial, and the notion of the two downing tools and strolling into the forest was more than he could bear. He took the gunner's grey cloak, young Greene's coveted possession, and presented it to Bylot, the first mate, with a ceremonial speech about Bylot's steadfast dedication to the task of finding the Northwest Passage.

Greene, on his return, challenged Hudson's decision.

"The cloak was mine," he raged. "It was not yours to give."

"I'll do what I like. Remember whose ship you are on, you ungrateful lout. It is I who allowed you to stay when you scrambled aboard in the night like a rat creeping up a line."

"But you gave me the cloak," protested Greene.

"Your friends wouldn't trust you with twenty shillings, and I was a fool to trust you at all. Let it be known to all that I hereby declare Henry Greene a stowaway who has come aboard our vessel to plunder and divide us. He will have no wages and must work diligently and obediently if he is to gain his passage home."

The crew was stunned. Greene had been, to that point, Hudson's favourite, suspected of receiving extra provisions, and certainly spending a great deal of time in the relative comfort of the captain's quarters. Greene had even boasted that the master intended to make him a bodyguard to the Prince of Wales when the *Discovery* returned to Britain. This was a complete change of heart, and the men didn't know what to make of it. Perhaps their leader was going mad. Inevitably, the seeds of mutiny were nurtured. Hudson, taking comfort now only in the presence of his son John, remained oblivious.

December came and went, and the deeper cold of January and February descended. All activity was forced to a halt as ice cracked the shrouds and sealed the hatches. The wind moaned and howled, sounding like a chorus of despair to the men in their bunks. They were down to their last rations. Even the ptarmigan and fish they had brought aboard in the fall had dwindled to nothing. Scurvy was pandemic, and men watched in horror as their gums blackened and they coughed their teeth into their handkerchiefs. Eating became an agony, and the pain in their joints was ceaseless.

Hudson divided the last of the food in mid-February. Each man received a pound of bread and a three-and-a-half-pound portion of mouldy cheese. Supplemented with moss scraped off the rocks and, by March, the occasional frog caught in the melt waters, the miserable company managed to survive.

As the spring sun brought lukewarm heat and the river began to flow, the water in the bay opened sufficiently to manoeuvre a small boat. Hudson, whose health had held up better than most of the others', decided to take the dinghy up the river in hopes of finding Natives who would trade meat in return for trinkets. But the task proved futile. The Natives had decided to shun the company of white invaders, to the point of setting fire to the woods when Hudson attempted to land his boat.

In his absence, Henry Greene and Robert Juet stole from bunk to bunk, enlisting all who had strength in their appalling plot to rid the *Discovery* not only of her captain, but of the sick men who were "eating up food that shouldn't be wasted on the dying." By the time of Hudson's return, seven mutineers had come on side, but they persuaded Greene and Juet to bide their time a while, as the ice was now breaking up.

Hudson weighed anchor on May 28, making his way out of the river estuary to the main water of the bay, where ice floes still posed a problem. By early June the *Discovery* was again trapped in ice, a predicament that led to Hudson's decision to remove Robert Bylot from the helm and replace him with an illiterate mate, John King. It was this action that finally incited the mutineers to action. On the night of June 18, 1611, Robert Juet and the bo'sun, William Wilson, assembled the seven conspirators and finalized their plan.

Early the next morning, as Hudson emerged from his cabin, he was grabbed from both sides by Wilson and Juet. Greene pulled the captain's hands behind his back and tied them.

"What manner of greeting is this?" bellowed the master. "Don't you know that your necks are at risk?"

"No more than your life," said Juet, who now held young John Hudson by his night-shirt. "Take your boy and over the side with you."

John King was also hustled to the rails, struggling mightily. The men and the boy were lowered into the dinghy. The carpenter, Staffe, declared that he would not stay on a ship run by traitors. He took to the small vessel voluntarily, but first persuaded the mutineers to leave him his chest of tools and a few items necessary for survival.

"Give us an iron pot and my fowling piece so we don't starve," he said, as the sickest men were dragged from their beds and forced into the dinghy.

The traitors complied. Abacuk Pricket, the sponsors' representative, was asked his choice in the matter: risk the gallows or face certain starvation. Pricket allowed as he would seek to justify their deed in England should they preserve his body and soul—which, to his mind, would be bowing to the will of God. Greene hesitated. He didn't like the snivelling ex-priest, but a quick head count made the mutineer realize that they needed hands enough to sail the ship to England. With nine men already in the dinghy, Pricket would be better off spared.

Bylot, who thought the men had been put over the side temporarily while the ship's stores were ransacked, was also spared. It was obvious that he must stay on board and alive if the *Discovery* were to make her way back to the Atlantic.

And so the deed was done. Henry Greene, quickly assuming the role of captain, cut the line that held the dinghy to the main ship. Hoisting the topsails, Bylot piloted the *Discovery* north through the spring ice as Henry Hudson and his small group of doomed men gradually receded until they were a mere speck in the distance. At last they vanished into the great undulating face of the sea.

What happened to Hudson and his castaways has never been determined. Some claim he survived for years on the shores of Hudson Bay; others say his fate was sealed within months. Relics of his company were found years later by Captain Tomas James, leading experts to believe that Hudson succumbed the winter following his abandonment.

Meanwhile, the *Discovery* was woefully undermanned, food was in short supply, and none aboard could agree on a course that would take them back to England. Bylot, who had been dubbed first mate by Greene, said their course lay to the northeast, but Juet insisted that the ship travel northwest in order to find the great capes that marked the channel. Bylot prevailed, with Greene's encouragement, and they sighted Digges Island some five weeks after the mutiny. Recalling the abundant wildfowl nesting on the island, Wilson and Pricket advised Greene to hove to.

Scouring Hudson's cabin after the mutiny, the men had turned up an emergency food supply—dried pork, cornmeal, peas, a small amount of butter—which sustained them for a while, but starvation was still nipping at their heels. The company was greatly relieved to take fresh meat aboard.

On July 28, a band of Inuit hunters was seen on the shore of the island. Greene decided to encourage trading, so as to ensure ample supplies were laid in for the long voyage across the Atlantic. Leaving Bylot and Pricket, who was still weakened by scurvy, to man the ship, Greene and four others went ashore in a pinnace. Greene had a weapon, but the others, expecting a friendly reception, were unarmed.

It is unlikely that these Inuit were the same that had encountered Frobisher decades earlier, but the story of that meeting may well have spread from band to band. In any case, the encounter on Digges Island erupted in a bloody battle. The Natives set upon them with arrows and hatchets. Greene took an arrow in the chest, dying instantly. In confusion and panic, the men scrambled for the boat and pushed off, leaving William Wilson fatally wounded and another man, Adrian Motter, stumbling through the waters in a desperate attempt to reach them.

"Take me up! Please, take me up!" he cried.

The air seemed thick with arrows, but a man called Pierce, who was rowing the boat, paused and lowered his hand, pulling Motter over the gunwales. The gesture cost him his life, for an arrow struck him in the back of the neck and he slumped over. Motter—with Juet, the only uninjured of the shore party—quickly took up Pierce's place, fearing the Natives would pursue them in their kayaks.

Bylot, in the *Discovery*, pulled the ship astern of the small boat. The two mortally wounded men, Wilson and Pierce, were hauled aboard, along with Juet and Motter. Pierce died within minutes, and Wilson, one of the original mutineers, two days later, cursing his fate as he breathed his last.

Of the original twenty-one men who set out from England, nine were left to sail the *Discovery* home. It was Robert Bylot who emerged the hero of the ill-fated voyage. Somehow, the navigator fought his way up Hudson Strait and steered the ship and her bedraggled, half-starved crew to the coast of County Cork, Ireland.

One man died en route. Robert Juet, the source of many of Hudson's trials, perished in the mid-Atlantic. Save for Abacuk Pricket, who felt compelled to utter a prayer, no one said a word or shed a tear as they weighted Juet's feet and pushed his body over the rail.

By September 1611, the *Discovery* was at the mouth of the Thames. Pricket and Bylot reported directly to their sponsors in London and were immediately thrown into prison on charges of mutiny. After consultation with the authorities, the sponsors decided that Bylot, who had almost single-handedly brought back the charts and notations of Hudson's voyage, should not be charged in the plot to overthrow his captain.

In 1618 Pricket, the surgeon Edward Wilson, and two of the original crew were brought to trial on charges of murder instead of mutiny. But with no body and no direct evidence that Hudson was dead, the men were eventually acquitted.

Bylot was given a different task. By way of punishment or reward, depending on your point of view, he was back aboard the *Discovery* in less than a year. Under the command of Captain Thomas Button and the explorer William Baffin, Bylot was again to sail north by northwest to search for poor Hudson and keep alive the elusive dream of a passage to the East. He was one of the only Englishmen at the time to have survived a winter on the frozen coast of Hudson Bay. He knew, better than most, that the chances of finding his captain alive were slim. But still he sailed, for Hudson's belief in the passage had been passed along to him. And his freedom depended on it.

JENS MUNK

A Deadly Winter

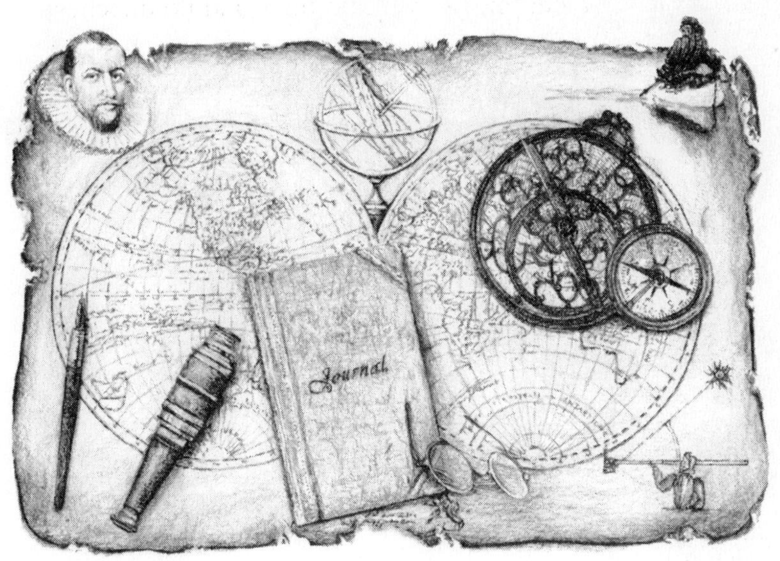

It wasn't the steady drip of the spring thaw that roused him. Nor was it the creaking of the mast, or the flap of canvas that sent him crawling on hands and knees up to the top decks of the ship. It was the stink of rotting flesh. It pushed Jens Munk from the foul nest of his bed below to the top deck of the *Unicorn.*

The six crewmen had all frozen shortly after they died. Their faces still held the grimace of pain, the mask of fear, the horror each experienced as he slipped from life into death. The thawing corpses were collapsing in on themselves, and the pungent, sickly smell of decomposition overpowered his senses and nearly accomplished what scurvy and the long, bitter winter had been unable to do.

Haltingly, weak as a new-born kitten, he inched his way up to the daylight. His tongue was swollen and parched, his face a dull grey, dazed and expressionless. He could not walk, and so dragged himself across the deck of the *Unicorn,* looking blankly at the ship he had so optimistically boarded less then a year ago. Much of the deck was still frozen. The ice and puddled water competing for space on the broad wooden planks made crawling awkward. He bent and sucked some melt water off the deck. It tasted sweet, and how soothing it was on his cracked lips and his toothless gums, gone mushy from lack of vitamin C.

The air was fresh. Its cleansing sharpness filled his throat and lungs. He turned his face to the sky, savouring the breeze on his skin. The air and light of the Arctic afternoon gave him a new lease on life. For Jens Munk, the Danish explorer who had followed Henry Hudson into the great bay in 1619 in hopes of finding the Northwest Passage, believed himself to be the only survivor of the past winter.

When the others had dropped like flies, one after another, he could only watch and pray that his own death would be gentle. Now this air, this sun—answered prayers, indeed. It was a weak spring sun, low in the horizon, but Munk knew it would melt the ice, the ice would turn to water, and he would be released from this horrifying place of imprisonment and death.

Eyes closed, head flung back, the memories of death and the smell of decomposing bodies dissipating in the spring air, the sound that escaped his lips was yet more a sigh of relief than a groan of pain. Then the miracle occurred: Munk heard a responding call, a shout from the land. He turned his head to the noise, sharply, ignoring the pain that rushed up his neck in protest at the rapid movement. Somewhere on that brutal shore, another soul had survived.

He saw a flash of colour against the snow-covered landscape. Focussing with difficulty after weeks in the darkness below decks, he made out two men sitting in a rough shelter under a tree, a small fire smoking before them. Monk turned back to his ship, tears of sorrow and thanksgiving streaming down his hollow cheeks. Once again, and for reasons he could not understand, his life had been spared.

The tale of Jens Munk is one of the most wretched in the chronicles of hopelessness and loss of life that make up so much of the history of Arctic exploration. In the winter of 1620, sixty-one men and boys died on the shores of Hudson Bay. Like Frobisher and Hudson before them, they had hoped to find the Northwest Passage. And, like Frobisher and Hudson, they encountered not a channel that ran to the Pacific, but ice and snow, sickness and death.

The town of Churchill, Manitoba, now stands near the site of the camp that became a graveyard on the banks of the Churchill River for the crew of two Danish ships, the *Unicorn* and the *Lamprey*. In 1619, when it became apparent that the company would have to spend the winter in the land the captain called Nova Dania, Jens Munk had dubbed his anchorage Munk's Haven. By the time the Canadian winter had finished with them, it might more aptly have been called Munk's Hell.

Born in 1579, Jens Munk was the unlucky second son of unlucky parents. His father, Erik, for reasons of his own, had refused to marry Munk's mother, and a child born out of wedlock at the time of the Danish Reformation was considered a child of the devil. When the bastard child was only twelve years old, his father was thrown in prison for misappropriating the King's money, and subsequently committed suicide. Thus, young Jens bore a double stigma: not only was he a bastard, but he was the bastard of a weak and iniquitous father. With no means of support for himself or his mother, the lad took to the sea, signing himself on as a cabin boy on whatever Dutch merchant vessel would have him.

Munk's youthful voyages took him throughout the charted world, and gave him an education he could never have attained in the classrooms of Scandinavia. It was with uncommon foresight that the young man jumped ship in Portugal and spent a year learning the language of Europe's primary sea traders. His fluency in Portuguese gave him a distinct advantage over his peers, and as he learned the craft of able-bodied seaman, he became a valuable asset on any missions of trade and exploration.

The young Munk lived something of a charmed life. Different accounts of his early years tell of him being plucked from the sea off the coast of Brazil following a battle between the Dutch vessel he manned and a French pirate ship. Munk and five others survived the shark-infested waters only to be deposited by their rescuers on the uninhabited shore of the Brazilian rainforest. There, starvation and disease claimed the lives of four of the six. The other two fought their way through the jungle to the port of Bahia, where Munk managed to keep himself alive until he was able to ship home with another Dutch vessel in 1598.

By 1603, after making his way from Amsterdam to Copenhagen and hiring himself out on a number of trading voyages, Munk had saved enough money to buy his own ship. The dream of winning the favour of the King and restoring the family name fuelled him, and in 1615 he outfitted his ship and sailed to the northeast, to Russia's famed White Sea, to get in on the fur trade.

He would have done well, except, like the British explorers who came before him, Munk was greedy for recognition. He sailed too far north too late in the season and was trapped by ice in the Kara Sea, the same place Hudson had been threatened by mutiny some eight years earlier. Abandoning his vessel and instructing his men to wrap themselves in all the furs they could carry, Munk led his crew across the ice to the shores of the White Sea, where he eventually talked a whaling vessel into taking him and his crew back to Copenhagen. His ship and his money were lost, but the resilient thirty-five-year-old refused to give in to defeat.

As he had hoped, his reputation as a master mariner came to the attention of the Danish King, Christian IV. Like Munk, the King had aspirations—not only for himself as the ruling monarch, but for his country, and, more particularly, his sailing fleet. He dreamed of ruling the northern seas, of taxing all vessels that dared to venture into the rich fishing grounds of the northeast. Having already laid claim to Greenland, Iceland, and the Faroes, Christian IV could see Denmark, in allegiance with Norway, one day controlling the wealth of all the Arctic waters. Before that was possible, however, they must find the dual gateways to the far Eastern market, the Northwest and the Northeast Passages.

The English were searching. Robert Bylot and his pilot William Baffin had thrice been to the northwestern waters, ostensibly to look for the lost Captain Hudson. In the process, the English had charted new waters and sailed to 76 degrees north latitude, beyond west coastal Greenland's

Melville Bay, to a place they had named Smith Sound. Lancaster and Jones Sound, on either side of Devon Island—north even of Baffin Island—were also now appearing on charts. According to the rumours, the English were well into the game, and if they were not careful, the mighty Danes, the original seafaring race, might be left behind.

In mid-January 1619, King Christian IV called a meeting with Jens Munk. He commanded the mariner to look to the northwest for the passage through the polar seas.

"You have prevailed against the ice in the eastern sea and found no way through," he said to Munk.

"This is true," came the response.

"Yet, you believe there is a passage lying to the northwest?"

"I do, indeed, Your Majesty. I have a chart of an expedition led by a man named Hudson. He—"

"Hudson!" roared the King. "Has he been found?"

"No, Sir, but his navigational charts show an almost certain passage to the East. I believe this is the direction we must sail, if Denmark is to secure the route."

"What do you need for this excursion?"

"A ship, Your Majesty, well victualled, and funds to pay Danish sailors a fair wage."

"Consider it done. The frigate *Unicorn* is at your disposal, and the single-masted sloop *Lamprey* might also prove worthy of this adventure. You took her from the Swedes?"

"Yes."

"Then that ship shall be yours as well," and King

Christian dismissed Munk with a few choice words of advice: "Choose sturdy, God-fearing men of the sea, Munk, and find for Denmark the Northwest Passage."

It was the commission Munk had been dreaming of.

On May 9, 1619, two well-provisioned ships weighed anchor from Copenhagen harbour, the forty-ton *Unicorn* with forty-eight men, and the smaller *Lamprey* with sixteen.

The expedition got off to an inauspicious start. Two weeks off the European coast, Munk was patrolling the decks when a sailor in front of him, without warning, threw himself over the side. It was an apparent suicide, but the incident deeply affected Munk. He retreated to his cabin, perhaps brooding on the similar demise of his own father. The loss of a man so early in the voyage and for no discernible reason cast a pall over the *Unicorn*. It wasn't until the southern cape of Greenland, named Cape Farewell by the English, was sighted on June 20 that spirits rose.

Forty days after leaving Denmark, the two ships were at the entrance to Frobisher Bay, mistakenly thinking it was Hudson Strait. The weather had been fair to this point, a typical Arctic summer with long, cold nights and brief periods of intense heat. Munk recorded in his journal:

> *On July 9 there was such a fog in the night, and so great a cold, that icicles were hanging from the rigging six inches long, and none of the men could stand the cold. But before three o'clock on the same day the sun was shining so hotly that the men threw off their overcoats, and some of them their jackets as well.*

Munk and his men were determined to endure, and found their way out of Frobisher's blind inlet in a matter of a week. They sailed past Resolution Island and up the north shore of Hudson Strait. Navigating with incomplete charts, the *Lamprey* and the *Unicorn* sailed southwest until they could no longer navigate the ice pack in the strait. Munk decided the best course would be to down sails, rope the ships together, and depend on the mercy of God to keep them safe.

The ships drifted through the ice floes. Pressure was a constant worry as the floating islands pressed in on their hulls. It was the *Unicorn* that eventually failed when a large piece of the stern, called the knee, which was bolted in place with six heavy iron rods, gave way. Water gushed into the hold. The captain ordered all carpenters to the stern. Working frantically, three ship's carpenters and two coopers tried to force the knee back into place, but their efforts were in vain. The ocean continued to flood in.

"We'll turn her 'round," said Munk, in sudden inspiration, "and let the ice press her back into place."

Calling on his men to work the rudder, he manoeuvred the *Unicorn* into an about-face. With the ice pressure now on the opposite side, the portion of ruptured planking was forced back into place and the carpenters were able to secure it with new bolts, adding cross planking for extra strength.

The refit was successful, but it served to remind Munk that as long as his ships remained among the ice floes they were in constant danger of sinking. He decided it would be prudent to look for a harbour with a decent anchorage, and stay put until most of the ice had cleared the strait. It was a wise decision.

Munk sent some men ashore to hunt for fresh game, and also to see if they could find any sign of human habitation. They brought back three birds. They reported seeing abandoned camps, but no people, it seemed, were currently living on the lonely shore. The following morning, however, the cabin boy rushed to Munk's quarters with the news that a number of men had gathered on the southern shore of the harbour.

"They are pointing at our ship and seem keen for us to meet with them," the lad said excitedly.

"Do they bear arms?" Munk inquired.

"It seems not," answered the boy.

"Then go to stores and take out the trading goods. Bring them to me, and we shall go ashore to meet the people of the ice."

The meeting with the Inuit turned out to be friendly, with Munk bestowing all manner of goods upon the people. The most popular item was a cheaply made hand mirror. The first Native man to look into the glass seemed confused and dropped the mirror immediately upon seeing his own countenance. It smashed on the hard ground, leaving fragments glinting in the sun. It was the first time the man had ever seen his face, except for the watery image reflected in the spring melt waters. The Natives were impressed, and laughter echoed across the water.

Through sign language, the Inuit indicated that they would return with gifts for the sailors. After carefully picking up every bit of glass from the rocks and wrapping the broken mirror in a rabbit hide, the strangers departed, leaving behind a seal pelt and some meat, and the promise that they would return. It was the last the European sailors saw of them. Speculating in his journal, Munk concluded that they were "doubtless subject to some authority which must have forbidden them to come to us again."

The two ships remained trapped by the ice until August 9, when the wind shifted and the strait was again free for sailing. A deviation into Ungava Bay lost them more time, however, and it wasn't until August 27 that the two ships completed the passage into Hudson Bay.

Rounding Digges Island, knowing he was more than a full month behind Hudson's schedule, Munk chose a southwesterly course. They sailed for days until a snow gale blew in, convincing the captain that they should again seek shelter near the shore. Separated from the *Lamprey* by the early blizzard, battling high seas and poor visibility, the *Unicorn* simply ran before the storm until land appeared to the south. It was the mouth of the Churchill River. Despite dangerous cross currents and rocks barely submerged in the heaving sea, the ship's pilot steered her into the inlet. Anchors were dropped, and the ship was in a safe harbour by the time the smaller *Lamprey* came into view, riding her anchors in the heavy seas.

"She'll never make this landing," warned William

Gordon, chief navigator of the *Unicorn*. "We'll have to show her the way."

"Take fuel, rags, and a barrel of tar ashore and make a blaze on that rocky headland," commanded Munk. "We'll send a boat to pilot her in. This is the only refuge, and we must stick together if we're to weather this storm."

The *Lamprey* was floundering. The rescue party was barely able to discern the top of the mast, although the ship lay in irons because the surf was so high. By some miracle, the crew were able to bring the sloop into safety, and, with the anchors fast, all they could do was hunker down and wait for the storm to pass. Munk's men, bone-weary and drenched, were just realizing that there would be no way out after the storm. The outlet to the southern seas, which Munk believed to be close at hand, would have to be found in the spring. It was now September. This was to be their winter resting place.

Munk had the ships pulled as high up the Churchill River as they would go, where they were beached. On the tidal flats he had his men dig a trench for the keels, and branches of trees packed with sand and clay were propped under the bilges to protect the ships from ice. The enterprising Dane further decided that action and a sense of purpose were needed if his men were to battle the depression that was the constant companion of Arctic expeditions. He drew up a roster of duties that would keep the men working as a team. Some were required to hunt for game while others scoured the land for dewberries, cloudberries, gooseberries, and any other edible fruit they could find beneath the skiff of snow that already lay over the land. A fire was to be kept burning on shore at all times in a makeshift open-air camp that the crew maintained. Snow had to be melted daily for fresh drinking water, and stones brought aboard for the construction of two fireplaces large enough to warm twenty men each.

These were built fore and aft of the mainmast in a sort of lounge area in which, it was hoped, warmth and companionship would ease the trials and fears brought by the winter. Munk had his men move the heavy cannons to shore to compensate for the weight of the fireplaces. Wood was cut from the dwarf timbers that grew on shore and, though the logs were small, they were ample. All the men moved from the *Lamprey* to the *Unicorn,* so only one galley was in use and only one ship needed be made winter ready.

The morning of September 12 dawned clear and sunny. Rising from sleep, the men were astounded at the sight of a huge white bear eating the flesh of a strange sea creature at the edge of the shore.

"It would feed us for days," remarked Munk to the assembled crew. "We must slaughter the beast and take his meat for sustenance and his fur for succour against the storms.

"While he is beautiful to behold," Munk continued when some men muttered their disapproval, "he would do more for the belly than the eye. Take muskets ashore, men, and we'll kill and eat the great white bear."

The men obeyed, although some were still concerned

about taking the majestic creature's life. Superstition and fear were not uncommon among seamen, and long after the deed was done, rumours circulated that killing the white bear would anger the spirits of the new land and cause hardship for the company. Their apprehensions were quelled with roast bear meat after the ship's cook first boiled the flesh then set it in vinegar overnight.

On October 7, with the fair weather holding, Munk decided to journey upriver. Taking a dinghy and two men, they rowed about a mile and a half inland before they were stopped by boulders protruding from the stream. Evidence of other humans was visible at the edge of the river: piles of chipped wood and cut timber, rings of stones where cooking fires had burned. On the way back to the ship, the men noticed some charcoal etchings in the likeness of the devil on a stone promontory. It spooked them enough that they christened the place Devil's Cape and hurried back downstream.

By the middle of October, the ice in the river was frozen and there was a hard frost on the ground. The hunters still went out daily, and though the yield was not as plentiful as in the autumn, it was still good. Ptarmigan and hares comprised the primary fresh food. Supplemented with a handful of berries and as much beer as the men wanted, it seemed to keep their health and spirits up. Munk was cautiously optimistic. The weather of mid-October in the Arctic was not unlike the coldest months in Denmark. They could not imagine anything colder, and blithely assumed that the worst was upon them. A month later, that assumption had been shattered. A penetrating frost cracked all the glass bottles aboard the *Unicorn,* and all the beer and wine froze, so that they had to be boiled in order to be drunk.

Inevitably, men began to fall sick. On December 12 one of two ship's surgeons, David Volske, died in his bunk. His corpse was left where it was, for it was too cold to leave the ship, and no one could get ashore to bury him. Not surprisingly, the sight of the dead man frozen in his bed troubled the sailors greatly. It was a grim foreshadowing of what was to come.

Munk tried to rally his company during the Christmas season. Roast hare was served on Christmas Day, and an extra ration of ale was given to all the men who had the strength to come to the galley. The pastor, a sickly stick of a man named Rasmus Jensen, gave a sermon extolling the benevolence of God and all there was to be thankful for. Jensen himself died before Epiphany.

On the shore of Hudson Bay in the frigid New Year of 1620, the vision of a passage to the East faded. The Danish sailors no longer spoke of tropical shores, of spices and gold, the riches to be won. The only thing they could hope for was the strength to survive the winter and the fortitude to somehow make their way home. Even Munk, who had suffered a season in the Kara Sea and staved off malaria on the coast of Brazil, was finding his faith flagging. The Northwest Passage was of little import now. His main goal was to stay alive and bring as many of his crew home as he could.

Scurvy had a different mandate. Caused by a deficiency of vitamin C, which is found primarily in fresh fruits and vegetables, the dreaded disease of mariners claimed the lives of thousands in the sixteenth and seventeenth centuries. Blackened and missing teeth, swollen gums, and mouth sores were the more obvious symptoms, but by the time a sailor was unable to take food by mouth, the joints in his body had turned to pulp, and death offered a welcome relief from pain. The berries they collected in the fall had helped delay the onset of the disease, but by January it was clear that it had commenced its deadly work among the crews of the *Unicorn* and the *Lamprey*. From January on, Munk recorded a new death almost every day.

Jens Munk begged advice from the remaining surgeon, Karl Casperson, who lay mortally ill. He thrust the ship's case of medicinal powders and remedies in the sick man's face. "Will any of these mixtures serve for the recovery—nay, even the comfort—of our men? I beg of you, Sir, stir yourself. The men are dying."

"And I, too," groaned the surgeon.

"You must help me. The labels on these vials are in Latin. I cannot make out one from the other. What must I do? Guess at a cure?"

"There is no cure, Jens. I have tried them all."

Munk grabbed him by his vest and jerked him upright. The medicine chest crashed to the floor.

"Help us!" commanded Munk. "You must help us!"

Casperson only moaned. He was beyond saving himself, let alone anyone else.

Munk released the physician and sank down on his

bunk, surrounded by the groans and odours of the dying. He looked back at the doctor in time to see the roving eyes halt, widen, then freeze in a death stare.

The days stretched into the future like a scroll unwound, recording the latest victims of the scourge. On January 23, mate Hans Brock died. On January 27, seaman Jens Helsing died. On February 5, seaman Laurids Bergen died. On the tenth, two more. On the twelfth, two more. The next day, another died, until, at the end of February, the dead counted over twenty and the sick numbered the same. There were now only seven people healthy enough to fetch wood and water.

As if dying in agony were not bad enough, disposing of the bodies was impossible. The living were too weak to dig graves or cart stones to mount on the bodies to keep wild animals at bay. There was nothing for it but to leave the souls where they lay. Their blankets were stripped from their bunks to warm the living, and the sleeping quarters below decks became a morgue in both fact and anticipation, as the ill and dying laboured for breath beside the stiff and frozen dead. Munk himself was suffering the effects of scurvy.

On March 4, there arose a glimmer of hope. The weather broke sufficiently for two of the crew to go hunting. They captured two hares, and Munk directed the cook to make broth of the bones so those in the last stages of scurvy would at least gain some small sustenance and comfort from the soup. As he took the common bowl and held it to the dying men's lips, he thought of the sacrament of the

Eucharist, the bread and the wine, and saw himself as the High Priest administering a last communion before death.

"We are in a poor way, and no one but God can save us now," Munk whispered. "What a melancholy people. Lord, where is thy mercy?"

March and April brought some relief from the cold, but no relief from the mounting death toll. The pilot, William Gordon, died on April 8, and Munk's lieutenant, Mauritz Stygge, gave up the ghost two days later. Good Friday, April 14, saw only four men other than Munk with the strength to rise from their beds. By Easter, forty-seven men were dead. By May 9, when it seemed certain spring would come, the body count had risen to fifty-seven.

On June 4, Munk was sure he would be the next to die. He confided to his journal:

> *Inasmuch as I have now no more hope of life in this world, I request for the sake of God, if any Christian men should happen to come here, that they will bury in the earth my poor body, together with the others which are found here, expecting their reward from God in Heaven. And, most gracious Lord and King (for every word that is found herein is altogether truthful) in order that my poor wife and children may obtain some benefit from my great distress and miserable death. Herewith, goodnight to all the world; and my soul in the hand of God.*

It was in early June, contrary to his expectations, that Munk was driven above decks by the stench of putrefaction.

The two men on shore, who had thought no one was left alive aboard the *Unicorn*, helped their weakened captain to shore. Crouched by the fire, the men could do nothing but gaze at one other, amazed that they had somehow been passed over by the Angel of Death.

It was spring growth that ultimately saved the men. They dug what green things they could find out of the earth, and sucked on the roots of the plants. As the ice on the river melted, Munk cast a net and caught six sea trout, a veritable feast after the deprivation of the winter.

Plucking bones from their remaining teeth, the survivors began to make plans for their return voyage. There were obviously not enough of them to bring home the *Unicorn*, but the little *Lamprey* might just be sailed with three men. As their strength returned, Munk and his two companions readied the ship. They had first to wait for the high spring tides to lift her from her winter berth on the flood plain. When, on June 26, they were able to float her alongside the *Unicorn*, the men began the grisly task of unloading the dead. The bodies were hauled up in sheets, wrapped and weighted, and unceremoniously tossed overboard. If there were prayers, they were silent. The smell made it impossible to do more.

After that task was complete, they provisioned the small ship from the stores of the larger one. The rigging on the *Lamprey* was refitted so the sloop might be sailed by the reduced crew. By July 16, the harbour was sufficiently ice-free to allow passage to the main bay. Before he left, Munk took an awl and drilled holes in the sides of the *Unicorn*.

"She will take on water this way with the rising tides," he explained, "and always be affixed to this harbour, this cursed place."

At last, the three were ready to sail. With a hasty prayer of thanksgiving and a stiff northwesterly breeze, they rounded the headlands of Jens Munk's bay and launched themselves into Hudson's black waters, making for Hudson Strait. Never far from danger, Munk and his men worked their way up Hudson Bay, fighting pack ice and unsettled weather conditions the whole way. They often went off course, but eventually, on August 14, Munk spotted Digges Island, the land that stood at the entrance to the channel between them and the North Atlantic.

Despite snow and sleet, the three men pressed on, focussing only on their own survival and their ability to keep the *Lamprey* afloat. While one pumped seawater from the hold, the other two attempted to avoid the massive drifts of ice that threatened to sink them. It was continual work; in order to get any rest, the men had to drop sails and let the ship drift.

As they rounded Cape Farewell, a gale blew up from the northwest, and for two full days the three men worked without rest to keep the *Lamprey* upright. When the storm had subsided, the bulwarks were battered, much of the rigging was in tatters, and both anchors lost. A second storm carried off the mainsail and mast. The beleaguered crew rode out the storm with a jury-rigged sail, and were all but ready to give up when they sighted land.

Somehow, the *Lamprey* had come within sight of the

Shetland Islands by September 13. Six days later, Munk and his two companions sailed their broken sloop into a bay off the coast of Norway. By courage, endurance, luck, and prayers they had persevered.

Sixty-three men had accompanied Munk on his mission to explore the Arctic waters; two returned with him. Instead of being welcomed by Christian IV and commended for his heroism and endurance in bringing home a ship, Munk was condemned for losing the *Unicorn* and cast into prison for two years. He was released after agreeing to mount a second expedition to recover the *Unicorn* and establish a Danish colony on the shores of Hudson Bay. Munk would have gone, too, had it not been for the intervention of the Thirty Years War. Munk could better serve his country as a naval commander than as an explorer, and Christian IV gave him command of a fleet of five ships including the refitted *Lamprey*. Mortally wounded in a sea battle in 1628, Jens Munk died at the age of forty-nine with the memory of his black winter and his abysmal failure still haunting him. He could only defy death so many times.

The Northwest Passage was no closer now than when he had first gone to look for it. Yet the human lives sacrificed to the discovery had been greatly increased by his adventure. The King, for his part, refused to acknowledge Munk's achievements or even attend his funeral, and Jens Munk went to his grave with neither fame nor the favour of the Danish Crown that he had fought so hard to gain.

FOXE AND JAMES

The Great Rivals

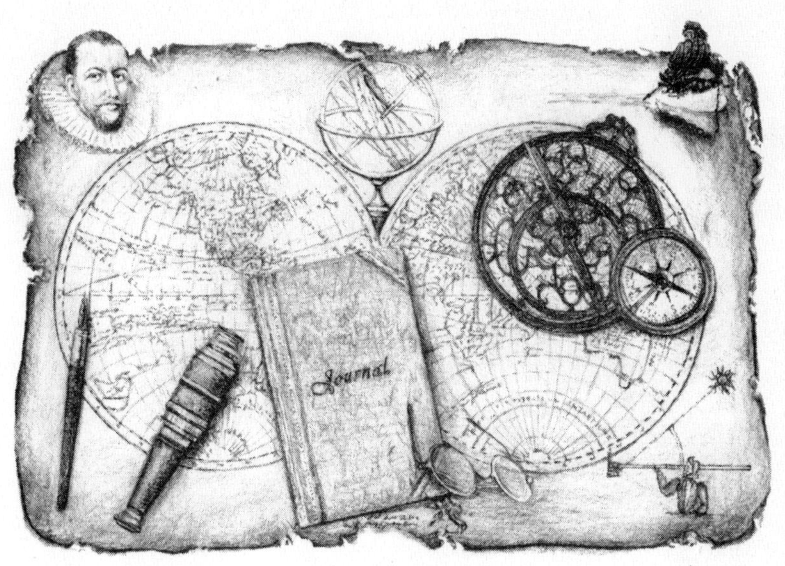

Rivalry, jealousy, a nation pitted against itself—such was the legacy of Luke Foxe and Thomas James as they each ventured into the icy waters of the Canadian Arctic in search of the Northwest Passage. Their stories could not be less alike, including the outcomes and conclusions of their different voyages. Nonetheless, the two great rivals cannot be separated in history or in narrative.

Luke Foxe, who had re-christened himself "Northwest" Foxe, went into retirement believing that the Northwest Passage would be found along the shore of Hudson Bay; an explorer need only search farther up that desolate coast to find the place where the land split in two to reveal the mighty waterway to Asia. Captain Thomas James, on the other hand, rightly surmised that there was no Northwest Passage from Hudson Bay—if, indeed, there was a Northwest Passage at all. If it did exist, he was convinced it lay much farther north of Hudson Strait and Baffin Island.

Two men, both Britons, both competent sailors, returned from simultaneous missions to the northwest with vastly different convictions based on what they had seen. The reason seems to lie in their experiences as much as in their different personalities. The dreamy Foxe, a Yorkshireman with a keen sense of adventure, spent a season poking around Hudson Bay before returning to London as the sailing season drew to a close and winter descended. James, a Welshman with an ordered mind and a calculating predisposition, over-wintered on the shores of the great bay and still managed to bring most of his crew home to Bristol. It was Thomas James who pointed future explorers farther north, cautioning them against entering Hudson Strait and Hudson Bay if they were serious about searching for the Northwest Passage.

Like Jens Munk before him, Captain James was deeply affected by the experience of a winter in the barrens. Unlike Foxe, he returned a skeptic, wondering if Britain and all the other countries that had fallen under the spell of easy trade with the East weren't just chasing their tails.

In some ways, both men were wrong. In other ways, each was correct. The passage *was* farther north, as James suggested, and it *did* exist, as Foxe believed. Foxe was the more optimistic, albeit misguided, explorer of the two, but it was his persistence that would encourage others to undertake the daunting journey of discovery.

Both men furthered the cause of the quest: James with his careful mapping, and Foxe with his undiminished dream. Together, their story is the last in early, seventeenth-century Arctic exploration.

Luke Foxe, born near the sea, had been obsessed with the notion of the Northwest Passage since he was a youngster. He would spend hours in bookstores and libraries and any venue that offered maps and charts of the world. He was a regular at the local globe-makers, and could often be found strolling the docks of London, engaging masters and mates and other seafaring folk in discourse about navigation, map-making, and the possibilities of finding the Northwest Passage.

In the spring of 1630, Foxe met a man of similar sentiments, a fellow Yorkshireman, Henry Briggs. Briggs, a noted mathematician and professor of astronomy at Oxford University, had a theory about the Northwest Passage. The two hit it off immediately.

"We know that, at the bottom of the bay, the tides are only two feet," explained Briggs, producing a hand-drawn map illustrating the Baffin/Bylot voyage of 1616. He pointed to the area south of the Churchill River where Munk had spent his disastrous winter, a place called Port Nelson. "Yet here"—Briggs drew his stout finger up the map some nine hundred kilometres northwest—"the tides are measured to be fifteen feet." He drew back in his chair and looked at the young sailor before him.

"Why do you think that is, Foxe?" He answered his own question before Foxe could respond: "It's the nearness of the South Seas, Foxe. The Northwest Passage! And you, Northwest Foxe, are the Yorkshireman who will sail it!"

The two men drew up a petition for King Charles I, requesting funding for an expedition. Before the petition was presented, however, Briggs and Foxe secured the support of Sir Thomas Roe, a diplomat with considerable influence in the royal courts. With Roe and Briggs behind the voyage, the King was reluctant to deny the request, and in 1631 he gave Captain Northwest Foxe the command of an aging navel vessel, the *Charles,* and the commission to "waste no time in the lower bay as Hudson and Button did, but stay north of Nottingham Island and follow the great current to the land of the Khan."

The *Charles* was a scummy, seventy-ton brute of a vessel. Undaunted, Foxe outfitted her with twenty strong men—many of whom had traversed northern seas before—two youngsters, and provisions for eighteen months. On May 5, with a stiff breeze blowing, the *Charles* left the Thames with Foxe as chief navigator and captain. The only blot on his optimistic horizon was that, at the request of Sir Thomas Roe, a government-appointed master, one George Dunne, was also aboard the *Charles*. Foxe despised him. He was a know-it-all and a troublemaker, and he certainly did not belong among the hand-picked professionals Foxe had engaged. Still, a voyage was a voyage, and rather than let Dunne spoil his delight on this voyage of a lifetime, Foxe decided simply to avoid the man. North by northwest the ship prevailed, around the tip of Greenland and into the waters that Frobisher, Hudson, Bylot, and Munk had sailed before them.

While Luke Foxe was approaching Hudson Strait, another British sailor was preparing to weigh anchor and make for the barrens northwest of Greenland. Captain Thomas James was backed by a quickly assembled group of businessmen from Bristol, the Company of Merchant Adventurers. These men were keen to send one of their own in search of the Northwest Passage. If London could launch a Yorkshireman on an Arctic adventure, they reasoned, surely the leaders of

Bristol could launch a Welshman. With little documentation as to why two similar missions set off from Britain within weeks of each other, spiting the Londoners remains the most likely explanation. And so the Company of Merchant Adventurers purchased the seventy-ton *Henrietta Maria,* hired a captain, and provided funds to employ a crew of twenty. In mid-June, a vessel only marginally sounder than the *Charles* set out from Bristol, a noted mathematician and learned scholar named Thomas James at the wheel.

Foxe and James had ships of the same size, weight, and vintage, and each carried an equal number of hands. But here many of the similarities of the two voyages end. Foxe's earlier start gave him a clearer passage through Hudson Strait and, despite the constant peril of icebergs, the *Charles* muddled through. The wind, mostly from the east and southeast, broke the floes into chunks, allowing the ship to chart a meandering course through the sheets of ice. Strong eddies and currents and the constant fending off of the ice floes kept the crew busy as the ship traversed the strait.

Equal to the turbulence of the seas was the clamour on board the *Charles.* Foxe had found it impossible to avoid the master, and their mutual enmity was now out in the open. Raised voices and quarrels frequently interrupted the work of the ship, and usually ended with one or the other of them storming off in a blue funk. Foxe was unhappy with the master and equally peeved at his mate. Both were lazy and easily distracted, and apparently unconcerned about the dangers posed by the ice. Halfway though the dangerous crossing of the strait, Dunne had approached Foxe with a request that the crew's liquor ration be increased. For Foxe, it was the last straw.

"Look around you, man!" he cried, indicating the great icebergs port and starboard. "These are cathedrals of ice—and flaked ice, too. These floes are deceptive and deep. We are in constant danger, and you want more liquor for the men? Are you mad? Each of us must keep our wits about us. We have to be vigilant and cautious. This is not a job for the faint-hearted wretch who must drown his fears in drink. Tell them no more liquor until we at least reach clear waters ahead. Do you understand me, Dunne?"

"All too well," came the reply. "But know you that troubles are at hand?"

"What do you mean by that, Dunne?"

The master scowled and lowered his voice. "I mean," he said mysteriously, "that this ship is doomed."

Foxe called a meeting mid-ship.

"I fully expect the ice to thin ahead," he told his crew. "We are near Hudson's great bay, and beyond is the open water that will lead us to the Northwest Passage. Take heart, good men. We are close to fair weather and calm seas."

Foxe's show of leadership seemed to cheer the crew, and they tackled their duties with renewed vigour. Beyond the strait was the bay, and beyond the bay . . . silk, spices, and glory.

Foxe proceeded westward along the northern shore of the strait closest to Baffin Island, despite exhortations and arguments from Dunne. Nearing Nottingham Island, which sits at the entrance to the channel, the watch from the mast-head called down the words that everyone had been anxiously waiting for: "Clear waters ahead."

A collective sigh of relief passed through the sailors. They had mastered the currents and the ice, and were now free to explore the northern half of Hudson Bay.

Going ashore with a few of his men at Nottingham Island, Foxe and his men found the remains of a shelter, but no sign of human habitation. A polar bear on an ice floe was killed the next day, and the fresh, gamy meat made a change from the cook's usual fare. It was mid-July, the weather was improving, and the treacherous strait was behind them. Spirits aboard the *Charles* were buoyant.

Sailing due west, Foxe probed a sound at the top of the bay, naming it Roe's Welcome Sound after his benefactor. When it proved not to be the hoped-for passage, he turned the ship south and sailed down the coast, laboriously recording all the inlets, sounds, bays, and rivers that flowed into the great Hudson Bay.

July slipped into August, and the days grew shorter. By August 9, Foxe was approaching Port Nelson, and he had still found no sign of the passage. The tidal flow was becoming shallow again, and he feared he had somehow missed the estuary of the great channel.

Going ashore near the confluence of the Nelson River with the bay, the company found signs that other British visitors had been there before them. A broken anchor, some iron fittings, and a tumbled-down cross marked the spot where Captain Thomas Button had over-wintered in 1612 following Hudson's disastrous voyage. Foxe had his men construct a pinnace, a smaller sailing vessel that could also be rowed, so exploration upriver could be undertaken. The boat was assembled and a large timber cut for the main mast, but again, inland exploration proved futile. None of the rivers flowed to China, none of the waters emptied into the Pacific. What they did discover were human footprints in the sand, a sign that he and his company were not alone. A tribe that Foxe guessed to be "nomadic forest dwellers" lived year-round on the great bay.

Back at the *Charles,* Foxe brooded. He had seen the geese flying south that afternoon, and at night the aurora borealis danced in the sky like the lost spirits of Hudson's crew—a sure sign that winter was approaching.

On August 29, as the *Charles* sailed south, an excited call came from the crow's nest: "Sail to the high side! Sailing ship, British insignia, two leagues on our starboard side!" And indeed, between Port Nelson and the deep gouge now called James Bay, on an expanse of water almost twice the size of the Baltic Sea, the *Henrietta Maria,* skippered by Captain James, met with Foxe's *Charles.* It was a bizarre coincidence—and, for both captains, something of an embarrassment.

Adhering to British manners, James sent his lieutenant over in a dinghy with an invitation for Foxe and the two men Foxe despised most, the master and his mate, to come for supper. It was an extraordinary, if somewhat strained, dinner party. Foxe was not a man to mince words, and wanted to know exactly what James had seen and done since arriving at Hudson Strait. Over roast hare on the top deck, with the *Henrietta Maria* rocking in the swell, Foxe posed his question: "And how did you find the straits?"

"Terrible," James replied, pushing back his chair and carefully wiping his moustache on a linen napkin. "We were at sixty-one degrees northeast when we heard a howl and a hideous noise that I took to be some shore beast or the breakers against that land, but nay, it was that awful, roaring entrance to the strait. The sound was waves crashing on a solid stretch of ice between us and the land. I first thought it impassable."

"Yes, yes," said Foxe eagerly. "Go on. How did you make it through?"

"With God's hand upon us," reflected James. "His mercy . . ."

Foxe, impatient for details, was not particularly interested in God's mercy, and interrupted his host rudely. "Yes, absolutely, but how was the sailing? Where did you find passage?"

"It was all ice, as well you know," said James, somewhat taken aback by his guest's manners. "We hauled her like a sledge from floe to floe, and finally anchored to two large ice packs that served as fenders and kept the larger floes at bay. It was a frightful way, for we were always in danger of being crushed by the ice. The battering was constant, and we made slow headway. We lost the kedge anchor when one of the floes suddenly broke in two, and, as my men went to retrieve it, we came perilously close to shore, drawing less than a few fathoms under the keel. I hoisted sail to drive us in toward the shore and we ran against a large plane of ice that held fast to the land. Still, the grounding drove in the main knee of the bow and broke the main shrouds to pieces. A good portion of my men, including the carpenter, who we now needed in great desperation, were lost in the pinnace and the fog had come."

"And what then? Were you taking on water?"

James nodded. "We were able to stay the flow, but there was more danger from without than within. The tides were up and the ice pounding against the ship. You should have heard the grinding! I thought we'd be sheered to pieces with the pressure."

"And . . . ?" Foxe prompted.

"At tide's ebb we were heeled over such that the rails were in the water. We'd worked all night securing cables from the mast to the rock on shore, but it seemed she was sure to go down. My men couldn't stand upright on the decks, so great was the pull, and I called to abandon ship before she sank."

"We prayed," James continued. "On the ice we knelt and prayed that God would save us. And"—he paused dramatically, sipping his wine—"our prayers were answered. God did save us. The tide turned and the *Henrietta Maria* was upright again. We sailed away from that place, Foxe. It was a place I've named The Harbour of God's Providence."

"And then you were through to the bay?"

"Yes, but not before a fortnight passed where we were caught again among the floes. The wind changed, and we worked our way forward with the men heaving her on, shoulders strong against the bulkhead. Once in the bay it was only to the southwest we could sail, and even then there was more ice than water as far as the eye could see."

Foxe nodded his head empathetically. He knew the dangers of the ice. But James had said nothing of the passage, the reason they were both here. Dare he ask? Yes, he must.

"I have a letter from His Majesty I am to deliver to the Emperor of Japan," said Foxe haughtily. "I hope to do so this season, before my return to England."

"I have found no way to China, if that is what you ask," said James. "Furthermore, I no longer believe that this is the way. Indeed, I wonder if there be a way to the East at all."

Foxe was incredulous. What could possibly have happened to so shake the man's faith?

James began to talk about a mathematical equation based on celestial navigation and the distance of stars one from the other. He talked of charting his course through the application of a mathematical manipulation called the Equation of Astronomical Time. Foxe, immediately and utterly lost in the verbal density of James's locution, quickly brought the conversation back to subjects he could understand. He spoke of the inadequacy of James's ship and the impossibility of over-wintering.

"You'd be mad to do it, James," he said.

"It is my duty," came the reply.

"What, to freeze to death in this foreign land? A colony of corpses won't further the English cause, my friend."

"But my commission is to stay and seek the passage, and while at times my faith fails, I have not completed my task. We will find a harbour for the winter and begin again this monstrous duty in the spring. You, Foxe, should do the same."

"The Lord frowns on those who take their own lives, James, and to rest on these shores is sure and certain death. But at least your bodies won't putrefy in the winter months. The cold air will stay that course."

There was enough truth in Foxe's reply that neither said anything more. They carried on with their meal until a reasonable time had elapsed and they could escape each other's company. James had found Foxe insufferable, cocky, and a coward to boot. Foxe had found James to be a fool, baffled by simple navigational skills and dependent on the jabbering theories of ineffectual magic men and pseudo-intellectuals who had never set foot on a ship nor charted the course of an expedition. Equation of Astronomical Time, indeed! Let the numbers save him when his ship was fast in the ice and the fittings cracking with the cold.

The men parted company on August 30, 1631, James carrying on a southeasterly course and Foxe diverting to the north. Enough of the shallow tides, Foxe thought. He would head across the bay again and up to the northern reaches to test Briggs's theory a second time. If he passed Nottingham Island on the western shore rather than by the east, and headed north, he reasoned, he would be sailing where no white man had sailed before. Let the pompous James continue southward. Foxe felt certain he was on the right course. *Those great tides must have recourse to another ocean,* he reasoned, and so, despite the reticence of his crew, the *Charles* sailed past Hudson Strait and into the waters between Baffin Island and the mainland, an area now called Foxe Basin.

It was a brave move on Foxe's part, but his passion for finding the Northwest Passage could have been tempered by a little more sensitivity to his men. They did not share his enthusiasm, and there was talk below decks about the oncoming winter and the fact that Hudson Strait might very well be frozen solid by the time their captain decided they should make for England.

George Dunne, who had never been on a serious Arctic expedition before, encouraged complaints and fuelled the dispiriting conversations, hoping they would have some effect on Foxe. He incited a group of men to break into the stores of liquor in the hold. When Foxe confronted the thieves, Dunne sided with the men, saying Foxe had "no idea of the hardship we suffer."

Foxe had read enough to know that a disgruntled crew could easily sabotage a voyage. With Dunne egging them on, and the frost already heavy on the scuppers, mutiny was a real possibility. As he saw it, he had little choice but to turn the *Charles* around and head back for Hudson Strait.

On October 1, after probing the waters of Foxe Basin and naming a number of capes and inlets after himself, Foxe came about and pointed the *Charles* toward home. He knew the season was over, and with many of his men freely helping themselves to open casks of sherry and wine, he had little control of his ship. Vowing never to speak to Master Dunne again, Fox steered into Hudson Strait on October 5. Battling heavy seas and the ever-present ice floes, he roused his men enough to get the ship through the dangerous stretch of water. Lashing additional canvas to the bottom of the sails, the bonnets were raised and the ship raced home.

Northwest Foxe had spent almost six months in Hudson Bay. He had sailed much of the west coast, charting it as he went. He had penetrated waters to the northwest where he was convinced the passage lay. But the weather and the temperament of his crew caught up to him, foiling his attempts to locate the exact latitude of the channel. Twenty-two men he had taken to sea; the same twenty-two returned. By any measure, it was a successful voyage, but Foxe would not be lauded as a hero on his return. Master Dunne and his mate maligned Foxe to his benefactors, calling into question his navigational skills and his ability to handle an unruly crew. Worse was to come, however, for in the court of King Charles, as Foxe gave his report, the monarch rebuffed him.

"I understand Captain Thomas James is still in my service aboard the *Henrietta Maria.* Our hopes to reach the Pacific now rest solely on his head. What say you to that, Foxe?"

What could Foxe reply? If the stubborn James didn't die of the cold, he would be squandering his sponsor's money while the ice prevented any exploration. Yet Foxe was the one being branded a coward for not over-wintering. In a fit of pique, he wrote a glowing account of his own journey, embellishing his accomplishments.

As it turned out, returning to London was a good decision. Although it earned him the reputation of a poor sport and a quitter, it did keep him and twenty-two others alive. Captain Thomas James, on the other hand, was in deep trouble.

Little more than a month after leaving Foxe, the *Henrietta Maria*, sprung with leaks from being battered against the rocks, her sails frozen solid, and trapped against the southern shore of James Bay by the encroaching pack ice, received news from the ship's surgeon that scurvy had been detected on board—and perhaps some strain of madness, too, for Captain James had been taken with the sudden notion that he would make his way overland to the "River of Canada," known today as the St. Lawrence. The weather soon returned him to his senses, however, and by mid-October he was forced to conclude that they were going nowhere. Despite blowing snow and bitter cold, he went ashore and scouted some eight kilometres of back country for game and berries to arrest the creeping illness that threatened his crew. But the takings were scant, and James returned to the ship downhearted. The crew, particularly those who were ill, requested that a house be built on shore, that they might feel the ground under their feet for their final days.

This pathetic request seemed to stir James, and he set a team of able-bodied men ashore to hew timbers and assemble a number of huts to shelter them for the winter. Meanwhile, the tide had turned and, rather than being battered against the shore, the *Henrietta Maria* was being pulled out to sea. Anchors attached to heavy cables were fastened to outcroppings on shore, but with the newly formed ice pressing against the hull, they were insufficient moorage against the tides and the Arctic winds.

As a desperate measure, James decided to sink the ship. The crew, not surprisingly, were incredulous, but James managed to convince them that it was the only way to preserve the *Henrietta Maria* and make sure they had a vessel to sail home in.

"She'll be dashed on the rocks or crushed by the ice, should we leave her as she sits," he explained. "What I propose is, we remove our provisions and that which we need to preserve body and soul, and take ourselves wholly ashore. When the wind should come northerly, our ship will either freeze in, or we'll stave in her sides and let her go to the bottom. If we don't, I fear there will be nothing left of the *Henrietta Maria* by spring: not the wood nor the iron fittings to make a pinnace and, therefore, no vessel to hasten us home."

"But if we sink her, how will we raise her again?" asked one of the men.

"We'll float barrels about her belly," James replied, "man the pumps, and trust the high tides of spring to take her up."

It was not a simple task. Emptying the ship of sheets and cables and sail and casks of hard tack and wine took innumerable trips to the shore. It was well into November, and the task not yet completed, when James decided they could wait no longer. The cold was deep and penetrating. The water was slush, thick with ice but not fully frozen, when the carpenter took his auger and drilled a hole the size of a small window into the hull near the keel. Leaving only sheeting between the hold and the frigid water, the men removed the last of the bread, gunpowder, and their few personal possessions. The final hour was at hand.

James forced his awl through the sheeting and let the seawater pour into his ship. It took hours for the vessel to settle, and the inside of the hold was battered by loose objects crashing against the bulkhead. The stern rudder broke away and disappeared in a swirl of water.

Slowly the *Henrietta Maria* gained weight and sank into the sands, until only the top deck and the mast were visible. It must have been a miserable moment for the sailors watching from the shore. Many of them might have wished themselves going down with the ship were it not for the shelters recently erected, and the promise of a cheerful fire inside each.

With the November winds whistling about the hut and snow already deep on the ground, the company of adventurers sat around their fire in the main hut with mugs of tea laced with rum.

"My masters and faithful companions," said James, "be not dismayed of any of these disasters, but let us put our whole trust in God. His will be done."

James continued: "If it be our fortunes to end our days

here, we are as near heaven as in England: but, in my judgement we are not yet so far past hope of returning into our native countries but that I see a fair way by which we may effect it. Admitting the ship be foundered—which, God forbid, I hope the best—yet have those of our own nation, and others, when they have been put to these extremities, even out of the wreck of their lost ship built them a pinnace, and recovered to their friends again."

"'Tis true," answered one sailor, "but I protest they were not in this position, with snow a full foot thick and mounting."

"It may be that they have happened into better climates," agreed James, "yet there is nothing too hard for courageous minds."

"Hear hear!" called out the carpenter.

"Hear hear!" echoed the crew, as each man's faith was strengthened in their community of hope. Captain James smiled. He had won them over. But would they, indeed, survive? Would their ship rise to take them home again? He bowed his head and led his men in a prayer of thanksgiving.

The next morning dawned clear and cold. James divided his crew into groups and assigned each duties. The first was to finish emptying the *Henrietta Maria* while the tide was at low ebb. The second was to row the goods ashore, and the third was to build a storehouse. Over the course of a few days, the men brought five hundred bundles of dried fish, the rest of the hard tack, and much of the bedding and clothing ashore.

The ice was temperamental. Some days a small boat

could be rowed out to the grounded vessel, other days the ice was thick enough that a man could walk across. When two men fell though the ice one day, James realized that time was against them. A cold snap on December 5 and 6 forced them to abandon the salvage operation. Five barrels of beef and pork and all their casks of beer were so solidly frozen into the hold of the ship that there was no way they could be shifted. The provisions on shore would have to suffice for the winter. The men were suffering from frostbite, and blisters appeared on their hands and faces as they tried to warm themselves around the fire. There was little to do but wait for winter to pass and try to stay alive. By Christmas, the symptoms of advanced scurvy were beginning to manifest themselves in James and his crew. Some had loose teeth and open sores in their mouths, their gums swollen. Others complained of swelling in their legs and aches in all their joints. By New Year 1632, more than two-thirds of the company was under the care of the ship's surgeon. But he, poor man, was at a loss.

"All my syrups and ointments are frozen, the bottles cracked and medicines shot through with shards of glass," he cried. "What are we to do to arrest this menace?"

James could only shake his head. The head carpenter, who had recently found a timber to shape into a keel for the pinnace that would take them out of this place, had succumbed to the sickness. The main hut, twenty feet square and hung with icicles, had become a sick room, with banks of bunks on the north and south walls. January and the long dark days of February seemed interminable as each man waited and prepared, in his own way, to die.

Miraculously, none of them did.

On March 15, one of the men, sitting in the doorway of a hut to avoid the smoke from green wood on the fire, thought he saw a deer moving about the woods near the storehouse. Three men went after the animal, but returned in the evening, hobbled by frozen feet. A fortnight went by before any of the three could stand again, and at one point they were all begging the surgeon to amputate their feet to rid them of the pain.

Easter came early that year. April 1 saw all twenty men still alive but only five of them healthy enough to work. After prayers and a special Easter scripture reading, the talk turned naturally to their coming voyage home.

William Cole, the carpenter who might have built them a pinnace, was too ill with scurvy to rise from his bed. The *Henrietta Maria* was still solid in the ice, but James thought it best to clear her at the first warm weather than try to cannibalize her to make a smaller ship that may not make it back though the straits of Hudson Bay.

With two iron bars and four shovel heads, the men began the painful process of digging their vessel out of her winter resting place. The decks were the first to be cleared, and once a fire had been made the prospect of getting back to the ship encouraged the men.

"Can we sleep aboard, Sir?" asked one of them hopefully. "It would save us the trek back and forth, and most certainly relieve us from the night howls and lamentations of the sick."

James agreed, and a camp was set up aboard ship for the half-dozen men healthy enough to work. A week after clearing the decks, the men penetrated the icy hold and, puncturing a cask, found it full of chilled beer.

By the end of April the pumps had been loosened from the ice and thawed, and the process of pumping out the hold began in earnest. By this time the sun was returning to rouse the frozen earth to life. It was too late for some. On May 6, the first of the sick men died. James called for the company to remove the corpse from the hut and carry it to the top of a hill. There, with a makeshift shovel and the words from the Service of Christian Burial, the men dug a grave in the unyielding earth and lay to rest the first casualty of the voyage. Two weeks later, the same group of men was on the same hill to bury the ill-fated carpenter beside his comrade. They had named the place Brandon Hill, after a landmark in Bristol.

"To last the wretched winter and die in the spring is the most mournful thing I have known," said James to his mate, as the body was lowered into the shallow grave. "William Cole was a fine Christian man and a better carpenter could not be found."

He shook his head and straightened the wooden crosses that marked each grave. Then he looked out onto the great inland sea that stretched before them, an eternal blanket of snow. But there, on the horizon, was a horizontal sliver of blue. Open water. Hudson Bay was breaking up. A shout of joy went up from the men, and they hastened to the shore to continue work on the *Henrietta Maria.*

A stroke of pure luck visited the camp in early June. A seaman by the name of David Hammon was idly poking his staff between the ice floes when he struck iron. It was the sunken rudder that had been sheared off when they scuttled the ship. The company was elated. Even the men in their sick-beds were cheered by the news. With a broth of beach weeds boiled in oil and vinegar, they were beginning to regain their strength. It looked as if no more men would be making the one-way journey up Brandon Hill.

Once the rudder was fished out of the depths, there remained the dual tasks of reattaching it to the stern of the ship and loosening the keel from the sand. Both projects could only be accomplished from the water, and six of the youngest and strongest crewmen plunged into the icy bay for six minutes at a time to clear away the sand at the bottom of the boat to create enough depth to hang the rudder on the stern post.

Six minutes was as long as it was safe to leave a human body in such cold. When the men surfaced, their lips were purplish blue and their skin had taken on the blue-white hue of a drowned man. As each diver emerged and scrambled back to land, the other men would immediately lay their hands on his skull to hasten the blood to his brain, otherwise he would faint. After six days the rudder was in place, ten tons of ballast had been heaved ashore, and the ship was ready to float.

Meanwhile, the hull had been repaired, and on June 22, 1632, the crew fastened ropes to their empty food and beer casks and forced lines under the keel. At high tide they laid their shoulders to the hull and pushed. The ship moved with glacial slowness through the sand. Still they pushed, drawing on reserves of strength they had not known they possessed. At last the water began to take her weight at the bows and the *Henrietta Maria* was afloat. Would the hull hold? Captain James held his breath, then slowly exhaled. His ship was riding high on the surf and little water was leaking in. God be praised!

James marked the occasion by felling a tree and fashioning a cross, to which he affixed a drawing of the King and Queen, sealed and edged in lead so it would weather well, and he erected a plaque that read: *Charles the First, King of England, Scotland, France and Ireland, and also of Newfoundland, and of these Territories, and to the Westward as far as Nova Albion, and to the Northward to the Latitude of 80 degrees, &c.* He also fastened the King's Arms, a shilling, and a sixpence to the marker. With no small pride and probably a fleeting thought to his rival Foxe, he also affixed the Arms of the City of Bristol to the cross. Then he and a company of men carried the cross to the height of Brandon Hill and planted it near the two grave markers.

As the morning of their departure grew near and the sick men grew stronger, the captain and a companion went up a high hill and lit a bonfire just to see if any humans might answer with a corresponding blaze. Unfortunately, the fire got out of control, and spread to the nearby trees and dry mosses. The two men scrambled back to camp, where they spent an unquiet night in the main hut, the air thick with wood smoke.

The next morning, with a sentry posted to watch the flames and the wind direction, a few frenzied hours were spent loading supplies onto the boat and rowing them out to the ship. Kegs of gunpowder, dried beef, and clothing were taken from the storehouse and the sleeping hut and hastened to the water's edge. The ship's mainsail, which had served as the roof of the hut, was unceremoniously removed and furled for transport. Shortly after noon, the sentry rushed into camp.

"Make haste!" he cried. "Make haste, the inferno is upon us! The winds have swung to the north and the flames are hard at my heels like a train of powder. Make haste. The fire is a mile across and our huts will be consumed."

He was right. As the last of the cargo was taken from the stores, the fire swept down on them, and the three small structures were reduced to ashes in no time. The men watched in awe from the beach as the fire continued toward them. But as they scrambled into the boat, and some jumped into the water, the wind again changed direction and the fire swerved to the east.

That night, James and his crew settled aboard the *Henrietta Maria*. The next morning, a contingent of the company rowed ashore for the last time. Ascending Brandon Hill, they observed a swath of charred timber twenty-five kilometres wide and reaching to the far horizon.

"Our Hell hath passed and all will be renewed," said James, bowing his head and saying a final prayer by the graves of his dead companions. He read a lengthy verse of rhyming couplets aloud, a tribute to the souls who had perished, and fastened it to the grave markers before turning away.

On July 2, 1632, Captain James cast off from their mooring at the bottom of James Bay. For six weeks the ship sailed through treacherous waters, making slow progress to the northeast. Each day brought what threatened to be certain death, so stormy were the seas. James was prepared to sink the vessel again rather than have her battered against the ice and rocks, but the crew would hear none of it. Instead, they filled their empty casks with water and stopped pumping the bilges in order to give the *Henrietta Maria* extra weight against the turbulent seas. Clearing themselves of the thick offshore ice, the ship was blown to the northwest by winds so fierce that Nottingham Island, the marker to Hudson Strait, was almost ten kilometres east of them when they finally reached the top of the bay.

Traversing the strait on August 26 was a marginally better experience than it had been the year before. Halfway through the passage, they struck a submerged ice pack and two metres of keel were torn away. Still, the vessel limped eastward, and on October 22, three and a half months after leaving James Bay, the ship arrived in the Roads of Bristol.

Captain James's report to the King was brief. Even if a Northwest Passage did exist, he said, which he highly doubted, it would be "very narrow and beset by ice and longer than the route to the East by the Cape." He himself would have no more of this sort of exploring. Presenting a roll of highly accurate maps to the King, he retired from active service on the sea. Less than three years later, on May 4, 1635, he died, disenchanted, adamant in the rather paranoiac view that his Northwest Passage expedition had been a fantasy conjured by rival traders who would have liked nothing better than to see the British founder and die on northern shores.

Luke Foxe, holding firmly to the belief that had sustained him through his life, cast about for command of another voyage to the northwest. There were none to be had. The English had other concerns as the seventeenth century matured and the aspirations of the Elizabethan era decayed. The search for the Northwest Passage came to a virtual standstill. For almost two hundred years, activity in the Arctic waters was restricted to French fur traders and Basque whalers.

But Foxe never gave up hope, and his final written words would prove prophetic. The way is there, he wrote, and whoever finds it "brings home the Golden Fleece."

Ross and Parry

Commerce and Conquest

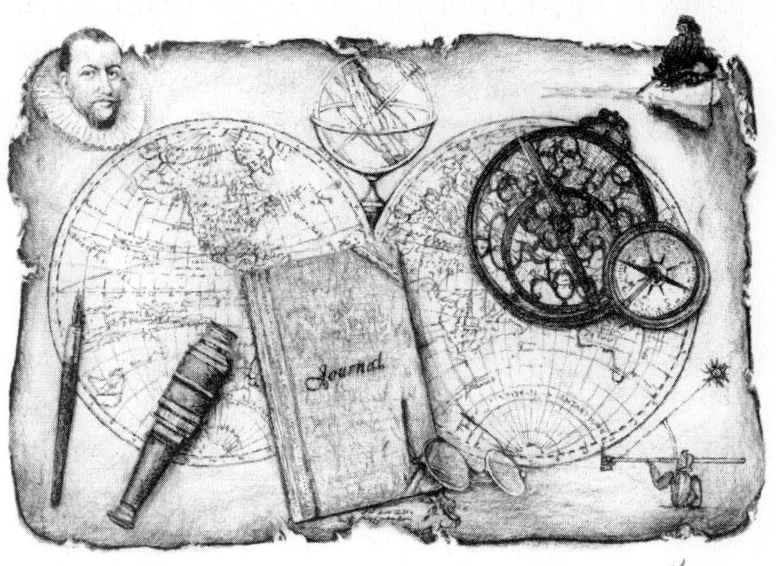

The snow-covered world emerging at the top of maps of British North America became a landscape of exploitation rather than exploration as the seventeenth century drew to a close. From 1668 until well into the next century, fur traders from French Canada were establishing Hudson's Bay posts along the shores of the bay that bore the name of the lost adventurer Henry Hudson. It was a time of conquest as French and English battled for dominion. The French subdued Newfoundland, while the English fought to gain a monopoly over the trading of beaver pelts from their far-flung northern posts.

The search for the Northwest Passage faded under the glare of money. Commerce usurped conquest, and profits became the sole reason for entering the icy waters of the northwest. With the notable exception of James Knight, a Hudson's Bay Company man whose ship and crew were lost in 1719 near Marble Island on Hudson Bay while searching for the Northwest Passage, there was limited exploration.

The ships and crews who penetrated Davis Strait on course for the passage were also pursuing commercial gain in the hunt for the great whale. By the end of the 1700s the Spitzbergen area was almost completely depleted of the mighty mammals, and whaling had moved to the western shores of Greenland and into Davis Strait. The whalers, of course, were up against the same harsh conditions faced by earlier explorers, and many died on the ice. But they didn't write books or command the attention of the rulers of the day as the northern explorers had a century earlier.

A new breed of adventurers came out of Britain as the eighteenth century faded. The imperial government of the early 1800s was not set on finding a route to China so much as leaving their stamp on the vast archipelago of northern islands that had been claimed for Britain by overland explorers such as Alexander Mackenzie and Samuel Hearne. The British navy had recently defeated Napoleon, but rather than disperse the fleet and put thousands of sailors out of work, the Admiralty again looked to the vast and, for the most part, still uncharted northwest. At best, they reasoned, they might find the Northwest Passage. At worst, a few men would die in the attempt, but at least the islands, the blank spaces on the maps, would be discovered and defined, and accurate charts bearing the British seal would secure English sovereignty over the Arctic lands of British North America.

Captain John Ross, the son of a Scottish village preacher, was the first to be commissioned in the quest for the Northwest Passage. In 1818 the First Lord of the Admiralty, Lord Melville, agreed to outfit an expedition for the purpose of exploration. Four ships were supplied.

The first two, under the command of a man named David Buchan, were to head north past Spitzbergen to the North Pole. The same route had been attempted by Hudson almost two hundred years earlier and proved futile, but contemporary observations of currents and birds and some wishful thinking convinced others of the existence of an accessible passage through the polar ice.

Ross's commission was different. He was given two vessels and charged to follow the current of water flowing from Davis Strait through Baffin Bay. He was to point his ship west as soon as the waters opened up, and, from the top of the world, sail through to the Pacific, anchoring briefly in the Bering Sea to drop off his navigational charts before re-supplying his ships in the Sandwich (Hawaiian) Islands. The benevolent but highly unrealistic Admiralty, deciding that Ross's voyage would take time—it was, after all, a couple of thousand nautical miles—granted him an entire sailing season.

Ross's ship, the refitted whaler *Isabella*, left Deptford on April 25, 1818. Behind the *Isabella* sailed the *Alexander* with Lieutenant William Edward Parry at the helm. Ross and Parry would know the hardships of their predecessors before the winter of 1818 was spent, yet both would push farther into the northern waters than any other white man before them. Like Foxe and James, though, theirs was also a chronicle of rivalry and betrayal, and where light and glory reigned down on one, the other would live under a cloud of almost universal disapproval.

Captain John Ross was the unquestioned leader of the first expedition. Parry was younger and less experienced, but he was a man marked by a different nature than that of Ross. The son of a physician, Parry had higher aspirations and a bigger ego. A man of charm and perhaps a slightly manipulative personality, Parry was happy to hang back and learn from Ross, noting both his accomplishments and his failings, on their first voyage. But he wasn't the type to play second fiddle for long.

Despite the perpetually dangerous conditions, August found the two ships well up the west coast of Greenland. They had made excellent progress, considering how many previous explorers had taken that much time to round Cape Farewell. But Ross and Parry had yet to feel the sting of the Arctic. The weather turned on August 18 as a gale sprang from the northwest. Churning snow and winds gusting up to one hundred kilometres an hour blew the ships back upon one another, until they collided. As if that weren't enough, an ice sheet the size of three football fields was bearing down on them. Through the snow, Ross thought he saw a channel.

"Get out on the ice, men!" he cried. "Take saws and make us a refuge dock before this floe crushes us altogether."

The crew jumped overboard and, with huge saws, desperately tried to cut a place for the ship where the bow had nosed into the channel. The ice was grinding against the hull, threatening to snap her at every turn, but carving a temporary harbour within the mass of ice proved impossible as the saws could not penetrate the depth of the floe.

"It's futile," cried one sailor to Ross, who had stayed at the wheel. "We'll have to let her go." Springing back on board, the men could only watch in wonder as the ice pressed the ship backward. Then, veering to port, it was swept away toward a range of grounded icebergs hard against the western shore of the strait.

"Bring her 'round!" Ross cried. "Run up the storm sails, and we'll run against the wind. This blow can't last forever, and God has seen fit to spare us with a shift in the wind's direction. It's our turn to work, men. All hands on deck."

Beating into the wind, the *Isabella* withstood the weather and stayed her course. Behind her, the *Alexander*, which had taken the brunt of the collision as the ice changed direction, was faltering but still afloat. Parry had watched Ross's vain efforts to carve a notch in the pack ice, but instead of attempting the same, had dropped sail and left his ship in irons to ride out the gale.

The sky was the colour of a bruised plum streaked with yellow when at last the blow abated, and Ross and Parry pressed on though water thick with floating ice. They had no other choice if they were to follow orders.

Slowly they made their way to Smith Sound at the top of Greenland to an extraordinary 78 degrees latitude. Eventually, forced back by an impenetrable wall of ice, the two ships veered to the southwest. As they sailed in a moderate breeze, an impossible sight was relayed from the crow's nest. There were men, dogs, and sleds ahead!

At first thinking them shipwrecked sailors, Ross hauled his ship about, positioning it so it could pull alongside the people. As they drew nearer, he realized that he was about to encounter the famed Eskimo of whom he had read. The captain, who had taken aboard a young Inuit man from

Greenland to act as translator to the Natives they would encounter, had the satisfaction of seeing his foresight rewarded.

He called John Sakoose to the bow as the ship came closer. Unfortunately, the party on the ice spoke a different dialect, and Sakoose had a hard time making himself understood. This, coupled with the obvious fear felt by the Natives, made the encounter strained. Eventually Sakoose was able to establish some preliminary details.

The Inuit were from farther north. They had never seen white men before. They addressed many of their questions to the wooden ship, because they thought it not an inanimate object but a living thing because of the flapping sails, which they mistook for wings. They were impossibly confused by the notion of anything other than ice and snow, for the world they inhabited was neither summer nor winter, but varying degrees of frost and thaw, always accompanied by snow in one form or another. A landscape with domesticated animals and forests did not exist for them, even in their imaginations. When Sakoose tried to explain the notion of England, the Natives questioned whether this place was on the sun or the moon. Which? After an exchange of gifts, the two cultures parted company, each to dwell on their encounter.

Pen-and-ink drawings by Sakoose depicting the exchange between Ross's expedition and the Inuit are on display at the British Museum in London. The contrast in costume is painfully evident in Sakoose's rendering of the scene. While the Natives are dressed in appropriate clothing of fur robes and sealskin boots, Ross and Parry are clad in regulation Royal Navy dress. On the cusp of the ice floes, it is easy to see why so many Europeans perished in their search for the Northwest Passage. Many made no attempt to adapt themselves to the ways of the Inuit, and non-adaptation in the climate of the north meant certain death.

After their encounter with the Inuit, the vessels continued their journey, entering Lancaster Sound toward the end of August. Keeping a westward course, the *Isabella* and the *Alexander* made their way through relatively calm waters. Excitement was mounting. Could this at last be the channel to the polar sea?

Parry was sure it was. Ross stifled his hopes. There was no distinct current in the channel and, as far as he could see, no evidence of driftwood on either shore. The mountains that bordered the inlet on both sides were consistently high, towering above the two small ships. There was no break in these hills, and the horizon did not promise a valley. Still the command was for forward sail.

On August 30, 1818, Ross thought he observed mountains at the top of the inlet. The channel had narrowed substantially, and it seemed that Lancaster Sound was to be yet another false start. Ignoring Parry's advice, Ross decided to turn the company around and sail back to open water. In his mind, the western passage was sealed. Parry disagreed, but could do little but obey his commander.

Brooding silently, Parry followed astern of the *Isabella*,

out of the sound and back into the waters of Baffin Bay. He vowed one day to return and probe the channel further. If it were indeed an inlet, he wanted to take his ship into its bay. Only then would he be satisfied that Captain John Ross had made a correct assessment.

It was understandable that Ross should choose to leave Lancaster Sound when he did. One of his astronomers was pressing him to explore Cumberland Sound, some 10 degrees latitude south of their current position. As well, the season had turned, and each night saw a drop in temperature. September was upon them, and they were far from the comforts of an English harbour.

After a disappointing foray into Cumberland Sound, the ships made for home. They reached the Shetland Islands by mid-October. Ross's first adventure was over, but Parry's career as a northern explorer was just beginning.

Back in England, Parry made a number of attacks on his commander, claiming that Ross had been simply too timid to explore Lancaster Sound fully. The Admiralty listened closely as Parry described the sound as "a most certain route to the Pacific." Ross, ever humble, admitted that, while the prospects of sailing to the bottom of the sound were poor, considering the mountain range he had seen descending from both northern and southern shores, Parry could be correct in assuming the inlet was navigable farther west than they had sailed. But what of it? Lancaster Sound was not the Holy Grail Parry believed it to be. The two were definetely at loggerheads, but the Admiralty chose to support the man with the more optimistic vision.

Parry was given the command of a second expedition, to set out in 1819. Its purpose would be to "explore the magnificent inlet" that leads to the Pacific. Ross was shut out. The British government wanted public opinion on their side, and they would rather mount a well-publicized voyage with specific intent rather than let a curious captain blunder through the ice, poking his prow into any potential waterway. The charismatic William Edward Parry, twenty-nine years old, darling of the popular press, was the man to lead the day. The steady and by no means flamboyant Captain John Ross was left to carry on his mariner's duties under a cloud of condescension and mistrust.

On May 4, 1819, Parry's expedition, consisting again of two ships—the 375-ton *Hecla* and the 180-ton, double-masted, square-rigged brig, the *Griper*—left Deptford for the Arctic. With Lieutenant Matthew Liddon as second-in-command aboard the *Griper*, Parry had ninety-four men under his command. One of these was James Ross, nephew to the disgraced John Ross.

Good fortune smiled on the company bound for Lancaster Sound, and Parry's first solo journey proved to be one of the most astonishing and successful in the history of Arctic exploration. Ice conditions that spring were unusually benign, and the ships were well into Lancaster Sound by the beginning of August. Innumerable whales, harbingers of open waters to the west, played in their wake.

Another thing cheered Parry. He detected a swell in the

channel. As they passed the point where Ross had turned tail, the young captain peered vainly into the distance to spy Ross's fictitious mountain range. There were mountains there, yes, but he was convinced they did not pose an impenetrable barrier.

On they sailed, with the *Hecla* in the lead and the *Griper* either abreast or astern. The question of the inlet's closure was answered once and for all when the mountains on the northern shore fell away and rounded out into a promontory, an immense bald cliff that opened to the south. A shout went up from the deck. The Northwest Passage was at last at hand.

Parry's jubilation was short-lived, however. Despite proving Ross wrong and charting Lancaster Sound as a passage to a larger body of water, he was disappointed to see a far-too-familiar barrier of ice in front of them. Except for the southern entrance into these uncharted waters, Prince Regent Inlet, the way appeared totally blocked by ice in the distance.

"Hard a-lee," bellowed Parry as the *Hecla* came about. Tacking through the open channel for 120 kilometres, the vessel came at last upon a wider channel, subsequently named Barrow Strait, which presented itself as a means of pursuing the western passage.

By then the ships were almost 650 kilometres from the eastern opening of Lancaster Sound. August was spent, September was upon them, still glorious but whispering of colder weather to come. In consultation with the master of the *Griper*, Parry found little cause to turn around.

"The ice is thin yet," he reasoned. "We've plenty of food, and it looks like these Pacific waters are as free from ice as those from which we came."

He was, in fact, gazing at a relatively ice-free expanse of Viscount Melville Sound.

"My vessel hasn't the weight of yours," cautioned Liddon. "Keep me always in sight and we'll continue our western course."

Parry was set upon finding the Northwest Passage, but he had other reasons for pressing forward. The British Parliament had offered a reward of £5,000 for the first ship to sail over the 110th meridian while still within the boundaries of the Arctic Circle. A two-day sail was all it took, and on September 4, 1819, Parry and company won the bounty by crossing 110° longitude at 74°, 44" latitude. The captain called for extra dried beef at dinner and a second cup of beer for his crew.

Again in consultation with Liddon, Parry decided it was time to find a safe harbour where they could over-winter. They decided to backtrack to an inlet they had passed to the east. The water was freezing at an alarming rate, and there was a distinct danger of Liddon's smaller vessel becoming immobilized far from shore. Even the heavier *Hecla* was finding it difficult to manoeuvre as the ice thickened.

On the south coast of Melville Island, now part of the Parry Islands Archipelago, the ships found refuge. But not before things became slightly desperate. To get close enough to land, the men had to carve a channel though the shore ice. The work was strenuous and time-consuming, for the ice was fifteen centimetres at its outer edge, thickening to as much as thirty centimetres the closer the men worked toward the sheltered bay. Carving a channel a mere hand's width wider than the *Hecla,* the crew worked with saws, hacking away rectangles of ice and either floating them out of the way or submerging them under the existing ice. They worked in four-hour shifts, with a half-hour rest in between, until the water below the main ship's keel was judged to be about nine fathoms.

It took four days of continuous effort to move the great ship and her mate into a safe harbour. The channel was over three kilometres long when the work was finished. But it was not a moment too soon. By the time the boats were winter ready, with their topmasts disassembled and their sails, anchors, and lines stashed on shore, the ice had already reclaimed the channel. The *Hecla* and the *Griper* were frozen in, side by side, and the long winter wait for spring had begun.

Mindful of keeping his men healthy, Parry had them build a snow porch against the port side of the vessel to store items that were usually kept on the top deck. This he cleared, that it might be used as a covered area for the men to take daily exercise. On a weekly basis, the ship's surgeon checked the crew's mouths for evidence of scurvy. But by November, when daylight was fleeting, no sign of the sailor's scourge had been noted. The company was in good health and well prepared to hunker down for an Arctic winter. November, too, saw the last of the great herds of caribou wander near the ship in their southern migration. Men were sent out with muskets to bring in as much fresh meat as possible, and both ships' stores were full to bursting as the temperature plummeted. As the crew entertained themselves with theatrical productions and the quieter amusements of cards, whittling, and story-telling, Parry, Liddon, and a few senior officers schemed about reaching the western end of the Northwest Passage by the end of the next sailing season.

The real winter, the endless, sunless season, came quickly, and buried the encampment in snow before crippling the movement of the men. Ropes were strung between the two vessels to ensure that movement from one to the other would remain possible in white-out conditions. This prevented the men from stumbling about blindly when they went from ship to ship. Twenty minutes outside was all it took for the cold to penetrate deeply enough to disorient the mind and persuade the body that it was best just to lie down and submit. Hypothermia was a constant danger.

A few weeks before Christmas, a small event occurred that turned out to have colossal implications. The glass bottles that held the *Hecla*'s supply of lemon juice burst in the cold. Six weeks later, the first signs of scurvy were noted in the physician's records. Parry plied the sick man with all the other anecdotes aboard—beer brewed from spruce needles, vegetable broths, and preserved fruits in sugar syrups. The symptoms fled.

Meanwhile, the ships were shrouded in the perpetual

darkness of an Arctic winter. Shunning depression and encouraging cheerfulness, Parry ordered each of the sailors to scale the mainmast in rotation for ten minutes a turn to see who would be the first to view the confusing twilight of mid-day which signalled the sun's return. It was on February 3, 1820, that a lookout first spotted the sun. The lengthening daylight beating back the darkness was enough to encourage the men to make plans for the spring and summer. The cruellest part of the year was past.

Parry and Liddon were anxious to continue their travels —so anxious, indeed, that they decided to venture overland and explore the icy heart of Melville Island before the ships were loosened from their frozen harbour. With a dozen men and supplies for three weeks, a small expedition trekked north. Their hope was that the central polar sea, which was believed to be a body of water that never froze, might be navigable. The group took lightweight carts with crude wooden wheels, and, harnessing the wind with blankets strung up as makeshift sails, they pushed and pulled these ungainly craft to the top of the bluffs and down again, only to find a sea as surely frozen as their own wintering grounds.

Back to camp they went, with nothing to do there but wait for the thaw. That year, unlike the two previous years, spring was cool and slow to come. The ice lingered, and, while the men could hear water gurgling below the pack and slosh through a good inch of melt water on top, the ships remained hard in the ice until the end of June. Another channel was cut, and by mid-July the company was again under sail. As soon as he was free of the ice, Parry insisted on heading farther west, but was almost immediately upon more ice.

He was not to know it, but Parry's expedition of 1820 came within four hundred kilometres of the western limits of the Arctic islands, those bordered by the Beaufort Sea. They had travelled farther than any European had before, and yet, by August, it was clear to captain and crew alike that they would not be able to sail through to the Eastern ocean without first returning to London to re-supply and refit the ships. On October 23, 1820, the *Hecla* and the *Griper* sighted the coast of Scotland.

Parry's adventures were by no means over. He knew the Northwest Passage existed, but he believed they were looking for it in the wrong place. How could one get past the ice that blocked Melville Island on three sides? How could one penetrate that great, immovable dome that capped the sea? What would happen if one sailed closer to the mainland, north from Hudson Bay through Thomas Roe's Welcome, the waterway that had been partially accessed but never fully explored by Luke Foxe two hundred years ago?

Between 1821 and 1825 Parry made two more voyages to the Arctic. Both gave definition to the evolving map of the Arctic islands and coastal mainland, but neither were as heroic or as dramatic as his first venture through Lancaster Sound. Over the course of three summers, Parry tested his theory of finding a passage closer to the mainland by sailing into Hudson Bay and then northwest. His ships, the faithful *Hecla* and the *Fury*, a ketch of similar tonnage, sailed to far western waters, where Parry was able to map the Melville Peninsula and discover a narrow channel that separated Baffin Island from the mainland, near present day Igloolik. He named the ice-clogged channel after his ships—the Fury and Hecla Strait—but numerous attempts to get into the channel were thwarted by whirling eddies and impossible ice conditions. If this was the way, it was barred to all but fools.

Parry's fourth voyage and third solo attempt to gain the Northwest Passage was again spurred by the notion that Arctic waters would be more open in areas where land masses prevented the formation of the huge sheets of thick ice he had witnessed in the northwestern reaches of Melville Sound near the Beaufort Sea and the impenetrable ice cap at the top of Baffin Bay. The passage, he felt, had to be within the islands.

In 1825, again with the *Fury* and the *Hecla,* he traversed Lancaster Sound for the third time and followed a southerly course down Prince Regent Inlet on the western shore of Somerset Island. His aim was to hug the mainland, but it was a bad year for ice. His ships were battered by the floes until the *Fury* was damaged so badly she could not be sailed home. Parry, by this time, was not only discouraged but tired. He decided to leave his broken ship on the shores of Somerset Island, on a beach in Cresswell Bay, thus leaving a cache of supplies and food for the next soul to seek the passage. Loading his men onto the *Hecla,* he left the Arctic islands for good.

He retired to England a heroic figure, the "Elder Statesman of the Arctic." As an advisor to the Admiralty, he basked in his reputation, not knowing that his first captain and mate, John Ross, would rise again to carry out the mission he had chosen to forsake.

In 1829, a full eleven years since he had made his mistaken pronouncement about Lancaster Sound, Ross sailed again. Unable to obtain support from the Admiralty, who had cooled to polar exploration on Parry's return, Ross persuaded a wealthy merchant and former mayor of London, Felix Booth, to underwrite the expedition. Ostensibly, his mission was to find the Magnetic North Pole, but written between the lines was the notion that his journey just might prove a passage out of Prince Regent Inlet to the waters of the western Arctic.

Taking his nephew on board the small and ill-equipped *Victory*, a 150-ton coastal vessel that had once run between the Isle of Man and Liverpool, Ross fitted the ship with an engine and a moveable paddle wheel that could be drawn up and out of the ice. His plan was to follow Parry's route into Prince Regent Inlet and head south for the mainland.

His nephew, James Clarke Ross, had been an able-bodied seaman and later officer under Parry, and he knew the perils of the seascape. More important, young Ross knew the location of the abandoned *Fury*, with her supplies safely stored under canvas nearby, and six boats drawn up on the shore. This experience and information were vital to Ross. He was carrying twenty-two men on a vessel considered small for the times, although it was larger and better equipped than the vessels Frobisher and Hudson first piloted into the northern seas. The auxiliary motor and fuel took up two-thirds of the space below decks. Ross needed to salvage the supplies of the *Fury* if he was to take his company beyond the final known point on Parry's 1825 map, Cape Garry.

Ross, now over fifty years old, was determined to vindicate himself, and the *Victory* set sail in April 1829. The passage around the southern tip of Greenland, Cape Farewell, was no more harrowing than usual. The Rosses, senior and junior, were up to the task, despite the fact that they had not sufficiently rigged their ship. The motor had proved worthless, the paddle wheel a joke. They were nothing more than ballast now, and the journey would have been much simpler and far less dangerous without them. Still, they sailed across the treacherous ice field of Baffin Bay and through Lancaster Sound, Ross marvelling at how easily he had been deceived by the blurring of mountain ranges into ice fog a decade earlier.

In his *Narrative of a Second Voyage in Search of the Northwest Passage*, Ross gives a telling account of the state of the seas at the time of his passage:

> *remember that ice is stone; a floating rock in the stream, a promontory or an island when aground, not less solid than if it were a land of granite. Then let them imagine, if they can, these mountains of crystal hurled through a narrow strait by a rapid tide; meeting,*

as mountains in motion would meet, with the noise of thunder, breaking from each other's precipices huge fragments, or rending each other asunder till, losing their former equilibrium, they fall over headlong, lifting the sea around in breakers, and whirling in on eddies; while the flatter fields of ice, forced against these masses, or against the rocks, by the wind and the stream, rise out of the sea till they fall back on themselves, adding to the indescribable commotion and noise which attends these occurrences.

It was in conditions such as these that the *Victory*, like a twig in a turbulent river, entered the deep throat of Prince Regent Inlet on August 11, 1829, following the course Parry had mapped out. Two days later, Ross and his men reached Fury Beach. They found no sign of the deserted vessel, but the much-needed provisions were in perfect order. Tinned meat, sugar, hard tack, flour, wine, and pickles were loaded aboard the *Victory*, as well as a good supply of gunpowder. Well-provisioned and in good spirits, Ross continued on his southern tack, keeping his ship close to the western shore of Somerset Island.

His luck did not hold. Between Somerset Island and the Boothia Peninsula is Bellot Strait, little more than a stream one and a half kilometres wide at its narrowest point. South of Barrow Strait, this is the only access to the western seas. Missing it, Ross would naturally conclude that Somerset Island was not an island but an extension of the British North American mainland.

It had been fair sailing for the *Victory* since leaving Fury

Beach, but by the time Ross rounded the lower cape of Somerset Island he appeared to be in a deep bay. The ice was piled up such that there seemed no break in the shoreline, and so the *Victory* continued her southern crawl, missing the channel that could have taken them between the two landfalls.

In mid-September, Ross had his men disassemble the useless motor and was searching for a spot to put it ashore. The temperature was plummeting, the ice floes multiplying, and the *Victory* increasingly in danger of shipwreck. Little more than three hundred kilometres south of Bellot Strait, halfway down the Gulf of Boothia—named after his benefactor, Felix Booth—Ross brought the ship into a protected harbour. Near the shores of the place he named Felix Harbour, close to the Boothia Peninsula narrows, the *Victory* was hauled through a channel in the thickening ice to her maximum draw of a little over two metres and left at anchor. Winter did the rest. By October 6 she was fast in the ice, and Ross and his crew were making preparations for winter.

It felt, he recorded in his journal, like "the prison door was shut upon us for the first time; leaving us feeling that we were as helpless as hopeless captives, for many a long and weary month to come." Little did John Ross know that this winter of 1829–30 would be the first of four he would weather in or near that godforsaken anchorage, and that the *Victory* would never be brought home to England victorious.

The vessel was unloaded and the motor cast on the beach. A school was organized for the days of monotony that lay ahead, with the elder Ross acting as chief instructor in the finer points of religion, navigation, and reading. A quick reckoning of the supplies confirmed that the company could last on full rations two years and ten months, such had been the bounty of the *Fury*. Heartened by their good fortune, the men built up a bank of snow to insulate the hull of the ship, and an ice-block house against her side to give them more room to move about. The upper deck was spread with snow and, as it packed down, sprinkled with gravel to make it passable. Thus, the entire ship was well insulated from the quickly descending cold.

Ross ran a tight ship. Despite having a year's supply of spirits, he declared the *Victory* an alcohol-free zone. Combining abstinence with Sunday sermons and daily lessons, they hunkered down to pass the dark days of winter together.

A group of Inuit visited the ice-bound *Victory* in January, the darkest month of the year. When they first appeared, stepping out from behind an iceberg, Ross was amazed. The Natives numbered over thirty, and they quickly surrounded Ross and the few men who had come to check out the rumour of fellow human beings. James Ross noted that each man was carrying a knife and a spear. He called out the word for greeting in an approximation of their tongue: "*Tima. Tima.*"

The Inuit answered similarly in a chorus of endorsement.

Ross called for his men to throw down their weapons, and they did so, again crying "*Tima*" to show that they meant no harm. The Inuit did likewise, and, returning the shout with a resounding "*Aja*," they opened their arms to demonstrate their defencelessness.

"This is good," said James to his uncle, and approached the men with open arms. Embraces were exchanged, with much laughter and apparent delight on the part of the Natives. While a runner was dispatched back to the ship to obtain small gifts of iron for the Inuit, Ross marvelled at their dress. Hooded parkas and trousers of animal hide were doubled over, with the soft hair of hide against the skin, the leather turned out and stitched to a similar hide. The feet were similarly clad, so that any melted snow or water would be repelled by the coat of the beast. They appeared healthy and well-fed. Although they were short in stature, their faces and bodies appeared plump. They smiled and laughed easily, particularly when presented with Ross's gifts of iron rings.

Calling the younger Ross forward, the Inuit attempted conversation with him.

"I think they are willing to accompany us back to the ship," explained James. "We should be honoured to host them."

The strange entourage proceeded across the ice to the *Victory*. Three men consented to come on board. The rest hung back, marvelling at the prow of the ship poking through the snow. The Inuit men were given preserved meat, which they ate without relish and tried to slip out of their mouths when their hosts weren't looking. They seemed to pity Ross's men for having to eat such poor food, and, while they tried to be polite, Ross noticed them shaking their heads sadly and muttering to each other in subdued tones. Far more interesting were the mirrors and candlesticks and lamps aboard the *Victory*. These provoked much glee among the Native men, who seemed to take particular enjoyment in holding the hand mirrors up to the faces of their companions.

Ross and his men enjoyed the company of the strangers, and after a second exchange of gifts, with the Inuit pressing on the white men a number of different types of bone and ivory tools, the two groups left each other with the promise that Ross would come to the Inuit village within the week.

On January 10, despite marrow-freezing temperatures of –38°C, a group of mariners, including both John and James Ross, travelled to the Inuit village, where twelve recently erected igloos testified to the ingenuity of the northern people. The Inuit showed off their soapstone oil- or blubber-burning lamps, and their sleeping platforms heaped with hides. The women and children of the village hung back, but John Ross noticed that all the females over the age of thirteen or so seemed to be married and living in their own quarters. Through guess and gesture, the two groups communicated about ice conditions, hunting grounds, and the return of the caribou. James Ross, who understood a smattering of the language, translated as best he could.

"They don't stay long in any one place. These huts," he explained, pointing to the igloos, "are built in a single day, so they can break camp at any time to follow the hunt. The caribou will be migrating north soon, and they will move south to meet the herd."

"Do they know of a channel from east to west?" asked Ross.

The Inuit drew an accurate map of the area, indicating no exit from Prince Regent Inlet. Ross was not entirely discouraged. A channel was mentioned—perhaps the one to the north, Bellot Strait, or maybe the Natives were speaking of Fury and Hecla Strait—but somewhere, when the season was right and the weather warm, a way would open to the Pacific. He knew it to be true, but if they were to continue their exploration, it seemed, they would have to send men out on foot to seek.

The Natives helped Ross and his company build sledges, and around April 26, with the first hint of spring in the air, James Ross and a party consisting of three Inuit, five of Ross's own company, and a pack of dogs to pull the sledges, set off across the Boothia Isthmus to the sea on the western side of the peninsula. They subsequently made several forays across the land, and each one provided new information to the expedition. In early May, James Ross showed his uncle a place where a channel lay, a kilometre and a half long and half as wide. It was full of water, but as yet it was only melt water. Underneath, the river was still solid, and under the ice, Ross suspected, was the gravel and barren soil of the Arctic islands. It was no place to drag a ship through.

On May 17, a further reconnaissance inland saw the company reach the western shore. They carried on out onto the sea ice, to the coast of King William Island via Matty

Island, and, heading north, traced the island to its most northerly point. Here, at Cape Felix, again named after the expedition's sponsor, Ross noted the quantity of ice on the shore. There were huge shelves of it, cast up on end as if the force that had put it there had considered it a task of little consequence. How could any ship compete with the ice? For there, at the northern limits of King William Island, was the butt-end of a phenomenon known as the polar ice stream. It was the same continuous and impassable blockage that had stopped Parry dead in his tracks in Melville Sound, just east of the Beaufort Sea.

Heading back to camp in a gale, James Ross was determined to know if this King William's land—he had no idea it was an island—was connected to the Boothia mainland. Like his uncle a decade before him, young Ross misread the telltale headlands as he looked south. He believed that he was looking into a closed bay, and that the promontories he could see were joined by a leg of land. They were not to know it, but they were actually looking out onto the frozen waters of the Northwest Passage. As the sledges struggled back to Felix Harbour and the relative safety of the *Victory*, the company was gratified, at least, that its overland expedition had broadened human knowledge of that previously unknown land.

It was time to take that knowledge back to London. Unfortunately, the slow spring of 1830 turned into a cold summer, and the *Victory* remained fast in the ice. In July, she was still paralyzed when the mosquitoes arrived in the millions and made the camp a living hell. John Ross was desperate to continue his journey. There were ample supplies, the men were keen to sail, and they felt they were tantalizingly close to an important discovery. Yet still the ice held.

It was not until September that it finally gave way. But September was too late, and new ice was forming at the *Victory*'s bows even as the wind filled her sails. Barely five kilometres north of their previous anchorage, the ship was pinned and sealed by the ice of a new winter. All through October the crew laboriously cut blocks of ice to create a channel that would enable them to drag their ship 260 metres into a safe harbour. Perhaps it would have been better if they had never left the original camp, but any activity now, Ross reasoned, was a welcome diversion from the certain monotony of a second winter.

The winter of 1831 was as cold and unrelenting as the previous one. Ross continually reminded the men of their ample supplies, the certainty of spring, the comforts of their ship. Desperate for anything to break the monotony, the men longed for the company of the nomadic Inuit, but none came. Instead, spring arrived. Again it was cool and blustery, but with the temperature finally high enough to venture out of the cramped quarters of the second deck, it was pure and joyful liberation. Knowing that the ice would hold the ship until at least the end of July, the captain made plans to take a number of his most restless men on a sledding expedition to find and mark the Magnetic North Pole, that roving northerly convergence that attracts all the magnetic compasses of the world.

At 96°, 46" longitude and 70°, 5" latitude, some 120 kilometres due west of their encampment, James Ross raised the Union Jack and claimed the Pole for King William IV. It was June 1, 1831—more than two years since the *Victory* had first been imprisoned by the ice of Prince Regent Inlet. But the temperature had risen to zero. A thaw was underfoot.

Unfortunately, it was not a complete thaw. In August a wind blew out of the west that drew the ice apart sufficiently that they could see a channel of blue leading out to the open waters beyond the bay. But the wind was fickle, and even with the *Victory* ready and waiting to move, it shifted and closed the channel. The company was facing yet another winter in the Arctic.

"They'll go mad, salivate, cast themselves out into the cold," raged James Ross to his uncle. "We can't continue. We must take action."

The elder Ross gazed at his impassioned nephew with empty eyes. "What choice do we have, James? What do you suggest we do?"

"We must abandon her."

"The *Victory*? And where are we to go?"

"Fury Beach," came the reply. "There are boats there, six of them, I believe. The men will not stay, and I can't say I blame them. We must take action, pack supplies, the chart, all of this, and take it with us. From Fury Beach, we can catch a whaler before it's too late. There's no other way."

The captain knew his nephew was right, and he began preparations to abandon the *Victory*. Even then, Ross refused to be rushed into an ill-conceived retreat from Felix Harbour that autumn. They faced a journey of more than 650 kilometres across a forbidding landscape. They would use that dreadful third winter to plan their escape, and leave in the spring of 1832. Had it not been for that hope, Ross and his men may have died of despondency, or the terrible fighting that erupts when men are caged too long together.

By April 1832, the company was ready to brave the elements and find their way overland to Fury Beach. An advance party went north on April 25, alternately carrying and dragging the boats from the *Victory* along the shore ice. Less than twenty-five kilometres from Felix Harbour they made their first supply camp. Back and forth they went, moving everything forward by degrees, struggling to survive outside in a spring that was colder than a mid-winter storm in northern England.

The waters of Prince Regent Inlet were not opening, so they abandoned the *Victory*'s boats at one of their base camps. A small group led by James Ross forged across the Bellot Strait toward Fury Beach. The others, under the leadership of the captain, struggled north with the bulk of the stores. But the news from Fury Beach was not encouraging. Three of the boats that Parry's last expedition had left on the beach had been washed away in a high tide. Of the three remaining, two had been damaged by exposure to the elements.

A month after abandoning the *Victory*, the company assembled on Fury Beach. A shelter was erected—a timber house of two rooms measuring 9½ x 5 metres—over which they laid a canvas roof. In Somerset House, as they called it, Ross and his officers planned their departure. The boats were strengthened; sails were cut to fit their masts. The provisions were divided among the boats. Everything was ready except the ice, which held through July. At last, on the first day of August 1832, a fissure appeared. It was enough to launch the boats into the open sea.

It took a month to manoeuvre the small, open craft up the coast to the southeastern tip of Somerset Island. Through freakish mid-summer storms, the sailors advanced only to find that Barrow Strait and Lancaster Sound were choked with ice. No whalers would be coming through there. They had no choice but to turn back for a fourth winter in the Arctic, this time in the dubious comfort of Somerset House.

Ross's men bore the signs of suffering and hunger, the ravages of frostbite, the patches of dead skin, unkempt beards, and matted hair. One can only imagine their state of mind. The Northwest Passage seemed a whimsical fantasy. Survival was the new reality. But with rations low and the first signs of scurvy appearing amongst them, survival, too, began to seem unlikely. Indeed, in that fourth winter in Somerset House, two men died of scurvy. Christmas Day, 1832, was marked by a meal of boiled fox carcasses. New Year's Day, 1833, dawned without light. The darkness prevailed until April, when James Ross and a party of men headed back down the coast to retrieve the boats they had abandoned on their initial journey to Fury Beach.

At last, on August 15, all was again assembled, and the

ice had finally parted enough to allow the tormented men to cast off once again. Ross said a prayer as his weakened men put their backs into the oars and rigged the sails for the attempt to reach the open waters beyond Lancaster Sound.

On the eleventh day they spotted a sail to the southeast, but the ship was soon lost between the swells as the wind carried her away. Before the men could feel the full weight of their disappointment, another cry went up from the leeward vessel.

"Sail-ho! Due north."

A ship was directly north of them. They watched in gratitude as a longboat was launched toward them. They were British sailors, Ross saw as the boat drew near. And something about the vessel looked familiar.

"What is the name and hail of your ship?" he asked the officer as the longboat came alongside.

"She's out of Hull, the *Isabella* by name," came the reply.

It was the same vessel that Ross had commanded fifteen years before.

The officer mentioned conversationally that the *Isabella* had once been skippered by Captain John Ross.

Ross nodded. "I am he."

"Ross? No, no," the sailor scoffed. "The man of whom I speak has been lost and is dead at least two years."

John Ross smiled. What did it matter if the man thought him dead and gone? His ordeal was over. Soon all England would know his name, and the fame that had evaded him in youth would finally come his way.

Sir John Franklin

Forever Lost

The loss of Sir John Franklin and 128 of the British navy's best has been shrouded in mystery for over 150 years. To this day no one knows exactly what happened to the two ships, HMS *Erebus* and *Terror,* as they bore Franklin on his third mission to the north in 1845. Theories abound. Many books have been written and scholarly papers published. All are rife with conjecture and hypothesis.

Sir John Franklin was lost. He is still lost. Neither his ships nor his body have ever been recovered. After his disappearance, a multitude of ships were launched to look for him. They were notably unsuccessful in their search for Franklin, but they did complete the charting of the coastal waters and the shore, and the demarcation of where a ship might thread its way around the Arctic islands eventually to cross from Atlantic to Pacific and back again. In effect, they finally found the Northwest Passage, although no ship was able to traverse it by sail or steam.

Franklin and his men comprised the largest and the last major sacrifice of human lives in the pursuit of the Northwest Passage. Was it worth it? As martyrs to exploration, the men of Franklin's third expedition, and especially the bold captain himself, loom large in the annals of Arctic adventure. Their story is one of tragedy, heartbreak, and a nation united in loss. But it is ultimately a story of victory—cold comfort for the men who perished, perhaps, but a triumph for humankind, whose 350-year quest for the

Northwest Passage ended with the return of the Franklin search fleet to harbour.

The outfitting of the *Erebus* and the *Terror* a decade after John Ross's miraculous return was the grandest of all polar expeditions. England was energized with thoughts of finally securing the Northwest Passage. Common people were taken with the adventures being played out in northern waters, and the country was alive with the notion that Britain was on the verge of a great discovery.

Much mapping had already been done by the time the Admiralty decided that a new expedition should go forward to penetrate the passage in 1845. Parry's observations showed a mere three hundred leagues of water between the eastern entrance to the Beaufort Sea, Bering Strait, and Melville Island. The fact that Parry reported a blanket of impenetrable ice between the two points did not deter Lord of the Admiralty Sir John Barrow in his enthusiasm to plot the Northwest Passage once and for all.

Had not the Rosses, John and James, been stranded on land within spitting distance—or at least a 280-kilometre hike south—of the same Arctic shoreline that a younger John Franklin had mapped in an overland expedition some twenty years earlier? Ross and company had trudged over King William Island five years later in 1832, and found their way out of Felix Harbour the following year. Yet, the map remained incomplete.

Sir John and the Admiralty decided that if a way was not found through Lancaster Sound and Barrow Strait to the Bering Sea, the expedition should head north through Wellington Channel, or southwest from Cornwallis Island, to find the westward channel and complete the partially drawn picture of the Arctic archipelago. The expansion of the British Empire was tantalizingly close.

The Royal Navy began to cast about for an appropriate commander for their expedition. James Ross, an obvious choice, declined the offer to sail north again, citing a promise he had made to his wife. Commander James Fitzjames was seriously considered for command after Ross refused but was turned down as too young. In the end, Fitzjames and Captain Francis Crozier, both experienced seamen and former Arctic and Antarctic adventurers, agreed to accompany the ill-fated expedition that would be commanded by Sir John Franklin.

Franklin was by no means unknown. He had been the thirty-two-year-old second-in-command on Captain David Buchan's failed 1818 journey into the Spitzbergen seas at the same time John Ross and Edward Parry had probed Lancaster Sound. A year later, Franklin had led an overland expedition from the shores of Hudson Bay to the fur trading post of Fort Providence on Great Slave Lake, following the route of Samuel Hearne. From there, he and a band of some fifty men proceeded up the Yellowknife River to Winter Lake, inland from the mouth of the Coppermine River, where they built Fort Enterprise during the autumn of 1820.

They over-wintered in Fort Enterprise, and in the spring a smaller contingent of twenty men headed further east, mapping the coastline of the Arctic Ocean. But they misjudged the weather, and left too little time for their return. At a place called Point Turnagain, almost 450 kilometres east of Fort Enterprise, Franklin and his men retraced the coast to Hood River and cut overland in a desperate attempt to find shelter against the oncoming winter. Nine men died of starvation on that march across the barren lands, the worst Arctic losses since James Knight and his crew perished on Marble Island in Hudson Bay one hundred years earlier. Unlike his predecessors Alexander Mackenzie and Samuel Hearne, Franklin was a babe in the woods when it came to overland travel in the Arctic. He was dependent on the skills of Native guides and hunters, particularly a leader of the Yellowknife band, Akaitcho, without whom Franklin and his entire company would surely have died.

Despite his minimal knowledge of subsistence living in northern climes, Franklin came back to England and convinced the British authorities to finance a second overland expedition to the same area, where, he reported excitedly, he had seen open water off the shores of the Arctic Ocean. Because his description corresponded with Parry's theory that the Northwest Passage lay between the mainland and the southern islands of the Arctic, Franklin was sent forth again to map more of the area.

This second northern expedition saw 640 kilometres of the eastern Arctic shoreline mapped, and John Franklin metamorphosed into Sir John Franklin, the darling of Britain's maritime society. His eleven (in addition to the nine who starved, one was murdered and one executed) lost men and his ill-prepared first adventure were forgiven and forgotten. With the nation's gratitude and blessing, he sailed down to the Adriatic Sea and spent a few years on duty in the Mediterranean and Greece. He then returned to Britain and was appointed colonial governor of Van Diemen's Land (Tasmania) for the next seven years. He was just shy of his fifty-ninth birthday when he returned to England, keen to head what he believed to be the final assault upon the ice. The Admiralty had found their man.

His appointment as commander of the expedition was announced in February 1845. Fitzjames was to be second-in-command on the larger ship, the 370-ton *Erebus,* and Crozier, who had sailed under Parry, was appointed captain of the slightly smaller vessel, the 340-ton *Terror.* Once the crew was in place, attention turned to outfitting the ships, both of which had been to the Antarctic under James Ross in 1843.

They were three-masted sailing rigs, sturdy navy vessels with reinforced hulls for breaking ice. But knowledge had advanced since they had been to the Antarctic in 1839, and the *Erebus* and *Terror* would be fitted out with all the modern technology. Two fifteen-ton steam locomotives, each with twenty-horsepower engines, were adapted to fit into the holds of the ships to serve as emergency power. The bows were planked with sheet metal, and a machine that would draw the salt out of sea water was installed on both galley stoves.

Fuel and food for three years were loaded, including hundreds of kilos of tea, chocolate, liquor, flour, wine, tobacco, candles, and soap. To keep scurvy at bay, 4,200 kilograms of lemon juice were rolled on board in massive wooden casks. The first contingent of eight thousand tins of food, an innovative new way of preservation recently and enthusiastically embraced by the navy, were also brought aboard. The contractor for the goods, a merchant named Stephen Goldner, promised more tinned meats, vegetables, and soups would be forthcoming, but his resources were overwhelmed by the size of the order, and he was falling behind.

By Monday, May 19, everything was in place. HMS *Erebus* had a library of 1,700 books, shelved and secured. HMS *Terror* had a further 1,200 volumes, including everything from scholarly scientific journals to the latest issues of the satirical *Punch* magazine. Should they become iced in, Franklin and his men would have no lack of reading materials. Besides books, each vessel had mahogany writing desks well supplied with paper, pens, and inks, and a primer to teach reading and writing to those among the crew who wished to better themselves. A hand organ was installed in each of the mess decks to help dispel the gloom of the long Arctic twilight. Instruments for scientific and geographical studies were checked and double-checked before the ships cast off.

On the morning of May 19, 1845, the *Erebus* and the *Terror*, with 134 men aboard, sailed out of the Thames under the banner of the Union Jack and cheers of approval from the crowds gathered onshore.

There were three contacts with Franklin's ships following their departure from England: the first at the Orkney

Islands, then at a port o' call in Greenland—where the ships were supplied with fresh meat and five crewmen disembarked—and an encounter with two whaling boats in Baffin Bay toward the end of July. It was the last anyone heard from them.

After the July meeting with the whalers, all contact with the *Erebus* and the *Terror* was lost. Subsequent explorers and scholars have pieced together the route the ships must have taken. The evidence has been gathered painstakingly over 150 years: a naval button in the hands of an Inuit elder, an ankle bone bleached white on the edge of the sea, a cairn scattered with rusted tin cans and bones.

By the end of 1847, with no word of their favourite son, the Admiralty was getting nervous. Three vessels were dispatched on separate voyages to find Franklin and his crew. A man named Captain Henry Kellett was to approach the Bering Sea from the west at the place that would mark the completed Northwest Passage, for hopes that Franklin would yet burst through the icy sea were still high. Sir James Ross, who had turned down the initial expedition, was fitted with a ship to follow Franklin's original path through Lancaster Sound. A third party, under the joint leadership of John Richardson and John Rae, was sent down the Mackenzie River in search of the missing men.

All returned empty-handed. That Franklin might be already dead and his crew starving on the barren beaches of King William Island did not occur to the rescuers. Ross's four-year feat of survival was fresh in their minds, and

Franklin had been provisioned for that and more. What most people believed was that he had been beset by ice, and was even now lounging in some frozen harbour, reading and eating and planning for the spring.

Ross traversed Lancaster Sound, setting off cannons at intervals. Against terrible ice and weather conditions, he was able to bring his two ships, the *Enterprise* and the *Investigator*, into harbour at the top of Somerset Island at the entrance to Barrow Strait. There he over-wintered, sending out search parties on foot to the west and across the top of the island and down its eastern shore. Again, all returned empty-handed.

Ross had under his command two junior officers, Francis Leopold M'Clintock and Robert McClure. By 1850 McClure was commanding the *Investigator* on his own, pursuing the search from west to east, through the Bering Strait and into the Beaufort. Seven years later M'Clintock launched a separate expedition in the *Fox*, a steam yacht purchased and equipped by Lady Jane Franklin, who was still desperate to find traces of her husband.

Meanwhile, clues pointing to Franklin's route were emerging. The initial three search parties had covered almost two thousand kilometres of Arctic landscape, yet had still come up empty. But the government of the day was determined to continue the search. In 1850, five years after the *Erebus* and the *Terror* had left England, no fewer than eight vessels were plying Arctic waters. The Admiralty assured Franklin's widow that the lost adventurer would be found—

dead or alive, they could no longer say. But Franklin's whereabouts would be discovered.

What *was* discovered that winter were the grave markers of three of Franklin's men. This was the first sign that trouble had struck the voyage mere months after passing through Lancaster Sound. The graves lay on the beach at Beechey Island at the western entrance to Wellington Channel, between Devon and Cornwallis Islands. On the lip of a thin spit of land that separates Beechey from the larger mass of Devon Island, three headboards jutted from the scree.

The captain of HMS *Lady Franklin,* one William Penny, was the first to reach the site. He found the graves unnerving, not so much for their appearance but by the inscriptions chiselled in the stones.

"These were young men," he said, pointing first to one weather-worn marker, then the others. "Not gone twenty-one this lad. A mere twenty-five years old this one, and this, a sailor of thirty-two. These were men in their prime."

Edwin De Haven, an American searcher on hand at the grave sites, pointed out the dates. "They all died in the first winter, 1846. Could they have been of poor health before setting out, or do you think something may have happened to the food supply?"

The question was something no one wanted to consider.

Sweeping the area, the searchers came up with other signs of inhabitation, proving that Franklin had passed his first winter on the east side of Beechey Island. A large cairn was discovered, at its base hundreds of open and discarded tin cans. The remains of a storehouse and what may have been a small armoury were also discovered. Franklin had definitely been there, three young sailors had been interred there, but what course did the ships follow after leaving Beechey Island?

Like the explorers who had gone before them, the men who searched for Franklin were at the whim of the Arctic weather. Before the fleet could come to a consensus, or even decide how to apportion up the possible routes among them, the frigid weather had arrived. The winter of 1851 was cold and unforgiving, and the searchers had to stay put until a late thaw released them.

In the spring of 1851, and again in 1852, more expeditions sailed from England—some to search for Franklin still, but others now seeking information on ships that had disappeared while searching for the lost explorer.

It was agreed that the *Erebus* and the *Terror* had passed through Lancaster Sound in 1845 and, finding Barrow Strait clogged with ice, proceeded north through the unexplored Wellington Channel. Ice at the most northern latitude, 77 degrees, might have forced them back down along the west and south coasts of Cornwallis Island before coming upon their wintering harbour, Beechey Island.

Additional information came from an unlikely source. Dr. John Rae, who had set out to find the lost ships in 1847 via an overland route, was no longer officially looking for Franklin. He was surveying the Boothia Peninsula for the Hudson's Bay Company when, in 1854, he heard a strange tale from a group of Inuit hunters near Pelly Bay. They recalled meeting a large party of dying white men on King William Island. The men were trying to walk south to Great Fish River. They had lost their ship in the ice, the Inuit recounted, and they were weak and starving.

"Did they have sleds?" queried Rae.

"Yes," said the spokesman for the group, "and they carried the limbs of their dead as food."

Rae blinked back his shock. Cannibalism? Surely not. He demanded evidence of the encounter, and the Inuit produced a number of articles, including silver utensils from the *Terror,* one of them stamped with Francis Crozier's initials.

There was no question that the Inuit had come across British sailors from the Franklin expedition. They guessed the encounter was in July or August 1848. The group had been walking on the south coast of King William Island. They were weak, sick, and close to death. But why were they walking? What had happened to the ships? And why were the men starving? They had departed with provisions in abundance. And what was this appalling nonsense about mutilated corpses?

Rae wrote a letter to the Admiralty, telling them of the information he had been given. Unfortunately, by the time his report became known in England, the country was immersed in the Crimean War. Franklin and his missing men were old news. It was, after all, nine years since the vessels had left England. The London *Gazette,* the mouthpiece of Queen Victoria, had published a notice that winter confirming the death of the men on Her Majesty's ships the *Erebus* and the *Terror*. Considered to have "died in the Queen's service," the names of the 129 officers and crew were listed and acknowledged before that particular page of history was turned. Public sympathy rekindled momentarily, then died.

ROBERT McCLURE

Fast in the Ice

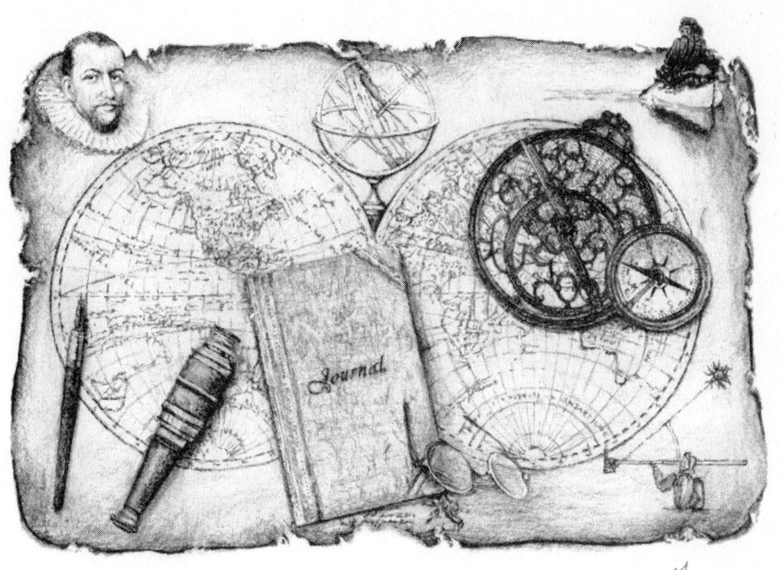

The first European actually to confirm the existence of the Northwest Passage and complete the arch at the top of the world was forty-six-year-old Robert McClure. He was the commander of HMS *Investigator,* a ship commissioned by the Admiralty to seek evidence of the Franklin expedition from west to east. It was McClure who walked from the northern shore of Banks Island across the ice to the westernmost point of Melville Island, Cape Providence—the site of Winter Harbour, the same cape visited by William Edward Parry on his first voyage through Lancaster Sound thirty years earlier.

McClure was certainly not planning to make the voyage by foot, nor did he have any knowledge of the grisly discovery of graves and bones that had been made in the eastern Arctic with regard to the men he sought. Unlike the other ships, McClure was on his own in the western Arctic. In 1850, the same year the three graves were discovered, Captain McClure had been commissioned to search for Franklin south of Melville Island, where the Admiralty still believed his ships were trapped in the ice. At the time McClure set out for Bering Strait, the search for Franklin was at its peak. If no signs had been found by the ships plying the waters of the eastern Arctic, it made sense to conclude that Sir John had continued west and was, even now, in the area Parry had eventually turned back from three decades ago.

McClure and Captain Richard Collinson, the commander of McClure's companion ship the *Enterprise,* sailed through the Strait of Magellan at the tip of South America in the summer of 1850. Emerging in the Pacific, they sailed up the western coast of two continents before becoming separated in a gale off the coast of Alaska. Although each believed the other ship to be in front, neither saw the other until their Arctic adventures were over some four years later. With sixty-five men aboard the *Investigator,* McClure sailed into the Beaufort Sea, convinced that Collinson had already passed the northernmost point of Alaska and was well on his way to finding the lost party.

There are varied reports of McClure's abilities as a sea captain. Some say that he was hard on his men, even tyrannical, that he was motivated primarily by ambition and greed, and ignored his sailors' well-being. Certainly he knew of the £20,000 reward to be dispensed by the government to "any party or parties, of any country, who shall render efficient assistance to the crews of the discovery ships under the command of Sir John Franklin." He likely knew of another £20, 000 reward, as well, offered to anyone who could ascertain the expedition's fate or bring relief to any of the crew. The information was common knowledge, though, and likely hastened more than one ship to the search zone.

Whatever the cause of McClure's zeal, it was rare for anyone to enter the Arctic Ocean alone, and the Beaufort sea was an uncommon point of entry for explorations. The majority of the water is covered by the polar ice cap, which never melts. A thin channel of open water follows the coast of the mainland from Point Barrow in Alaska to the mouth of the Mackenzie River and on to the islands that make up present-day Kitikmeot. But these waters are only navigable at the height of summer, and even then the pack ice is forever shifting. The threat that winds or currents will blow this solid mass against the shore is constant. It was through these perilous waters that McClure steered the *Investigator,* no doubt against the ardent wishes of his crew.

At the delta of the Mackenzie River, where fresh and salt water mixed, the ice had been pushed farther from shore. Here, between Banks Island and the mainland, as the days darkened down toward winter, McClure began to seek a harbour for his party. He pressed eastward through the wider channel of water now known as Amundsen Gulf until the land on his starboard side veered sharply northeast.

Between Banks Island and the Prince Albert Peninsula on Victoria Island was a narrow waterway leading to Viscount Melville Sound. It was the Prince of Wales Strait, a 250-kilometre gutter edged by cliffs of ice. It was the end of September by the time McClure reached the top of the strait. He had achieved Viscount Melville Sound, but, except for a narrow skirt of slush and icebergs around the islands, the main body of water was a mass of pack ice. Rather than retreat, McClure decided to let his ship drift just outside the pack until it became frozen in. It was a decision that almost cost McClure his ship, his crew, and his life. Mountains of ice, some three or four times higher than the *Investigator*'s main mast, drifted on the open water. The pressure on all sides buckled the deck and caused such a creaking and groaning in the hull that the men believed their vessel would split in two before the deep freeze of winter anchored her once and for all. Fortunately, the fickle winds came up just in time to force the *Investigator* partway back down the narrow strait. There, southwest of the channel opening, close to the shores of Banks Island, the ship was sealed by ice.

Once the company was organized and a routine established for the long, silent season, McClure picked six men to cross to the top of the island with him. It was nearing the end of October. Icicles formed on beards and moustaches. Exposed flesh froze in minutes and turned black within hours of thawing. But McClure was a man possessed, and was willing to take extraordinary risks. He pushed his men onward, pausing only when he reached his destination. There, gazing across Viscount Melville Sound to the squat blur of darkened shore that was Melville Island 110 kilometres away, McClure knew he was looking at the final link in the Northwest Passage.

Somewhere out there, he believed, was Sir John Franklin, frozen into the ice the same way he was caught. Next summer, he thought with satisfaction and determination, he would rescue Franklin and lead him back through Lancaster Sound and thus become the first person to sail the Northwest Passage from west to east. Exhilarated, anxious to get on with his purpose, it was a reluctant McClure who returned to the winter routine of the ship.

Spring, as the Inuit know, cannot be hastened. It was not until mid-April 1851 that McClure could again take up the search for Franklin. Three parties went out from the ship as soon as it was warm enough to venture above decks. Their mandate was to fan out and follow the shores of both Banks and Victoria Islands east and west, north and south, to see if any evidence of Franklin's expedition could be found.

McClure, still believing Franklin was farther northeast, was almost relieved when his men came back with reports of game and the stirring of the ice, but no signs of the lost expedition. The triumphant sail across the sound would complete the Northwest Passage, and on the other side, he was sure, Franklin would be awaiting his rescuers.

Toward the end of June the *Investigator* was ready to sail again. Heading to the top of Prince of Wales Strait, which thawed long before the pack ice in Viscount Melville Sound, McClure was again thwarted by sea ice. Determined that the ice would not triumph, McClure called his men to turn the ship around and head back the way they had come. They retreated down the strait, and on the western side of Banks Island they again steered north, approaching Viscount Melville Sound from a different direction.

There McClure found conditions as bad, if not worse. But he pressed on, for he saw it as his only option if he hoped to reach Viscount Melville Sound that summer. There was barely any navigable water. With a three-metre wall of ice on one side and white cliffs of gravel and ice on the other, the ship had little room to manoeuvre. She could only continue her dangerous course, with the yards alternately scraping the pack ice on one side and the land mass on the other. Had the wind shifted even slightly, the boat would have been crushed like an eggshell. His men were prepared at every moment to leap to the ice, where they would probably freeze to death, but at least they wouldn't go down with the ship.

The effects on the crew were beginning to show. Tempers were ragged. Men broke down and wept for no reason. McClure, sleep deprived and manic, teetered on the thin edge of sanity. Still they pressed on, sometimes sailing, sometimes hauling the ship from rowboats, sometimes pulling her forward as the men walked along the shores of ice, weaving between bergs and floes so large they dwarfed the ship. At last, in the third week of September, a deep bay opened out like a mirage to the east.

"We'll take her in," said McClure, to the relief of his comrades. "We'll stay through the winter. This is our Bay of Mercy."

And so the harbour was named. Had McClure known that his ship would never sail out of that perfect port, he might have christened it Bay of Horror, or Bay of Despondency.

The winter of 1851–52 passed without major incident.

The sixty-five men were cramped, perpetually cold, and some were sick in body or spirit, but these things were to be expected on a polar voyage. They lived as best they could in the belly of their ship, and as soon as the wind lost its chill, McClure organized a group of six men to accompany him across the great ice field to Parry's Winter Harbour on Melville Island.

Perhaps he expected to find Franklin and his men there, or at least a store of provisions. What he found, in fact, in the cairn erected on a point of land near Winter Harbour, was a letter from a Lieutenant M'Clintock informing whomever found the message that the vessels searching for Franklin had come and gone. There was not a sailing ship within five hundred kilometres.

McClure is said to have wept when he read the letter.

After leaving their own message, detailing their whereabouts to the nearest latitude and longitude, the disheartened company left. All they could do was go back to their ship and wait for the summer sun to free the *Investigator* from its Bay of Mercy. But in mid-August the ice still held. September came and went, and all hope of moving on was crushed. Supplies were growing short; rations were cut in half. Even so, some men started to decline food because of aching teeth and sore mouths. Scurvy was spreading on McClure's ship.

The black winter of 1852–53 saw an epidemic of torpor and depression descend on the ship. Men dragged themselves out of their bunks only to stare for hours at the blank white face of the indifferent world. Sleep was their only escape, and it brought little comfort when the ever-dropping temperatures left them shivering in their bunks under rough and inadequate blankets. McClure was as gloomy and unproductive as any of his crew. He hadn't the heart to rally them. As the winter dragged on, two men suffered nervous collapse. One laughed and howled uncontrollably, and the other curled up in his bunk like a child and refused to rise or communicate. McClure knew that the rest of them would either die or follow these unfortunate souls into the nether world if they spent another winter in the ice.

His thoughts turned to Captain John Ross, a fellow naval officer who had suffered four winter-bound seasons off Boothia Peninsula. Ross and his men had eventually fled their ship, risking an overland journey rather than staying with their immovable vessel one more season. And almost all of them survived. McClure would do the same.

Still, what if this were the year the ice pack loosened its grip? What if this approaching summer was the season they could free themselves and sail through the Northwest Passage? He hesitated. He decided to split the company into three groups. The strongest men would stay with the ship and sail her home should the thaw prove kind. The two other groups would take sledges and hike across the ice—one party to the east, making its way toward Lancaster Sound, and the other to the south, across Banks Island to the mouth of the Mackenzie River, where civilization of some stripe was almost certain to be found.

At the end of March 1853, he called his men together to tell them of his plan to escape the Bay of Mercy. They took the news mutely. They seemed not to have the energy even to discuss it among themselves. The men who were to be left with the ship were the strongest. The others were starving and stricken with scurvy. How could they possibly trek hundreds of kilometres in miserable conditions pulling sledges?

Still, they accepted their orders. It wouldn't matter, in the end, what group they were placed in. The odds of their getting out of the Arctic alive were diminishing rapidly. McClure was gambling with his crew's lives—they knew it and he knew it—but they also knew he had no other choice.

He chose April 15 as the departure date, but on April 6 the course of events changed dramatically. McClure was out on the ice, examining the hull of his ship and talking to one of his lieutenants about how to bury the body of a crew member who had passed away in the night, when a figure came into view, coming toward them across the ice. Supposing it at first to be one of his own men, McClure was not alarmed. As the figure approached, however, he began to wonder if the stranger might be an Inuit hunter, lost and seeking shelter.

The person was shouting something, but the words

were caught by the wind and carried away. McClure stood his ground. As the man drew closer, details sharpened. The stranger's face was black, as though he were African, but out of the ebony face shone two remarkably blue eyes. The black skin, McClure realized, was the result of frostbite. The man was a British sailor.

"I am Lieutenant Bedford Pym," he gasped as he staggered forward. He threw an arm back toward the direction he had come. "Her Majesty's ship, the *Resolute,* under the command of Captain Henry Kellett, awaits you on Melville Island."

One can only imagine the effect these words had on the weak and starving men aboard the *Investigator*. They were likely incapable of shouting for joy, but this was like a last-minute reprieve for men who had been sentenced to death.

Over a modest meal, Pym relayed the story of his own ship. The *Resolute* was part of the Franklin search squad, he explained, but they were also seeking McClure and Collinson. Two other ships, under the command of Captain Edward Belcher, had been launched north of Devon Island to look for them there. Together with her sister steamer HMS *Intrepid,* the *Resolute* had penetrated Barrow Strait in 1852 and managed to struggle through to Melville Island until they were trapped by ice. Because they were sixty-five kilometres short of Winter Harbour, Captain Kellett had sent a group of men up the coast to the cape to lay in provisions for the search that would begin in earnest that spring. The men returned to the ship in late

October, having found McClure's message earlier that year.

"'Twas too late in the season to send anyone across the ice for you," explained Pym, "but I volunteered to make the hike as soon as it was bearable. I left more than a fortnight ago and found my way by compass and sense of smell." The wiry officer laughed, a sound that hadn't been heard on the *Investigator* since the previous summer.

"What of Franklin?" McClure's response was curt. "Has he not been found?"

"Not as far as I know," said Pym. "Poor fellow. Not much hope for that lot. It's almost ten years now since he disappeared, and well you know what the time in the ice can do to a man's soul, let alone his body. Stiff, I think, Franklin's boys. Or sunk down in the water, frozen solid every one."

It wasn't the message McClure wanted to hear. He had defined himself as a rescuer, and here he was in need of rescue. As Pym spoke about the best way to get his ailing men across Viscount Melville Sound, McClure began to wonder if abandoning the *Investigator* was, in fact, the best course. Without a ship there was no way he could be the first to sail the Northwest Passage. And what would be the reward for an officer who returned to the Admiralty without his vessel?

His thinking changed the following day when the ship's surgeon reported that two more men had died of scurvy in the night. There was no recourse but to abandon ship and usher his beleaguered crew across the ice to Melville Island

and the relative comfort of Kellett's vessel. Lashing the most feeble men to sleds, the company bade farewell to the ice-bound *Investigator* at the end of May. Weary, hanging on to each other for strength, they struggled more than one hundred kilometres to Dealey Island where the *Intrepid* and the *Resolute* were still fast in the ice.

McClure and his men were offered warmth and food. The weakest among them were fortified with medicine and then taken with some of Kellett's men to Beechey Island, where a supply ship picked them up. They docked in England in mid-October 1853. Like John Ross before them, they were likened to Lazarus returned from the dead. All England was amazed at their survival, and the Admiralty, who had counted both McClure and Collinson among the missing, was now second-guessing its instructions to Captain Edward Belcher. Four ships had sailed into the Arctic in 1852. Some of the men they were supposed to be rescuing were now back in England, while the remainder—McClure and the more able-bodied among his crew—had remained with Captain Kellett aboard the *Resolute*. Belcher, however, was still looking for McClure, who had been looking for Franklin, but now that McClure had been accounted for, Belcher's whereabouts were unknown.

McClure and the men who stayed behind with the ships, however, were fortunate that Belcher and the second half of his flotilla were still in Arctic waters. A fourth winter was to pass before they were able to get out of Viscount Melville Sound.

Belcher, who had passed two years in Wellington Channel, was determined to avoid a third winter in the Arctic. He sent a message to Kellett and McClure in the spring of 1854 to abandon ship and rendezvous with him on Beechey Island. As far as Belcher was concerned, life and limb were far more important than ship and sail. And so the remnants of McClure's company, as well as a good portion of the crew of the *Resolute,* walked the hundreds of kilometres to Beechey Island, where they met up with Belcher, who had abandoned his two ships in Wellington Channel.

The men were loaded onto transport ships in Lancaster Sound, and all of them sailed for England in the fall of 1854. McClure immediately made it public that it was he who had first made his way from one ocean to the other, and he was therefore the true discoverer of the Northwest Passage. He was knighted and acclaimed. Belcher and Kellett, meanwhile, were court-martialled for abandoning their vessels. Although they were later acquitted, neither were to share in McClure's spoils.

After considerable debate in the British Parliament, McClure was awarded £10,000 for confirming the existence of the Northwest Passage. Whether the passage was commercially viable, or even navigable, was yet to be demonstrated, but the Admiralty had to admit that the channel that had consumed the imagination of Europe for the past three centuries had at last been verified.

Granted, it had not been sailed. It hadn't even been fully charted. But there was no question that Captain

Robert McClure had sailed into the Arctic from the west, crossed the great frozen sea on foot, and sailed out of the Arctic on the other side of the world. It was a distinct victory for Captain McClure. His financial award signified not only the existence of a Northwest Passage but also that Franklin, with all his splendid technology, was not the man to find it. Shortly after McClure's return to England, the search for Sir John Franklin was officially abandoned.

FRANCIS M'CLINTOCK

A Last Attempt

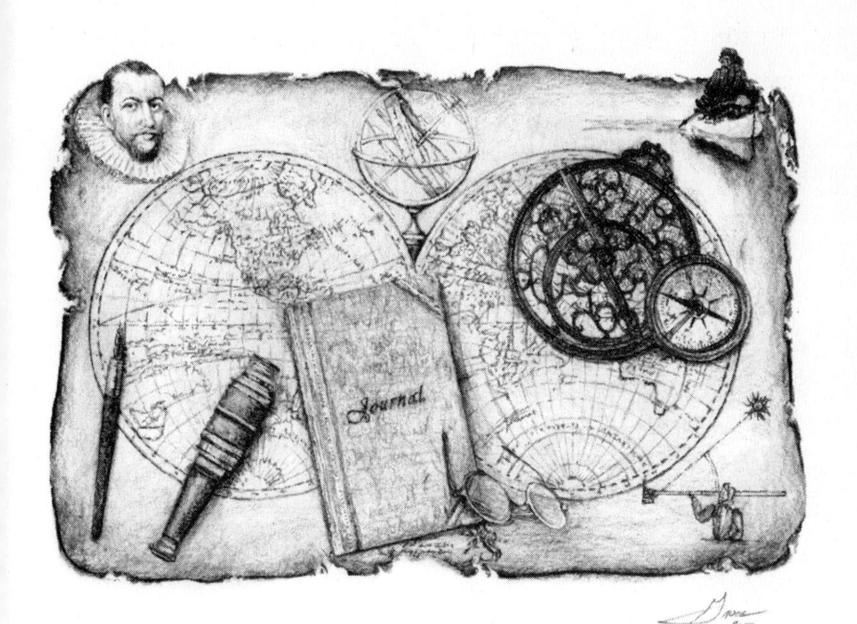

When the message came, Francis Leopold M'Clintock was not surprised. He had been half expecting it ever since he'd heard that the Admiralty had called off the search for Sir John Franklin. Lady Jane Franklin, with the help of public donations, had raised £3,000 to purchase and refit a luxury steam yacht, the *Fox*. Now she needed someone to sail the vessel northwest to discover once and for all what had happened to her beloved Sir John and the men entrusted to him. The request was something M'Clintock could not decline.

Lady Jane, born Jane Griffin, had been involved in the search for her husband since it first appeared that he had gone missing in 1847. She had since outfitted three expeditions, spending much of her personal fortune, and hounded the navy to continue the search when all hope failed. She believed passionately that Sir John would be found, and the rumours of starvation and cannibalism that were rife in Britain would be put to rest. The woman whom most of England thought of as Franklin's widow—a word she herself refused to speak—had an enormous influence on both public sympathy and the Admiralty. While ships were being outfitted and lost and abandoned and refitted to try again, Lady Jane became the personal face of the Franklin tragedy. As the years lengthened and the hope of Sir John's survival diminished, Lady Jane made sure that the search remained a parliamentary and political priority.

She became an expert in the topography and geological phenomena of the north. When she felt that interest in her husband's cause was waning, she made headlines by appealing to other nations for help. She unabashedly approached both the President of the United States and the Emperor of Russia, asking them to publicize Franklin's plight and join forces with the English in the search for her missing husband and his crew. Almost daily she went to the docks to quiz whaling captains on what they had seen or heard in the desolate regions that had claimed her lover. As her passion became an obsession, the public reacted with unfailing compassion and respect.

A woman ahead of her time, Lady Franklin was well travelled, well spoken, and well thought of in an era when wives were little more than necessary appendages to any respectable man who wanted to further his career. With few exceptions, women of her class in mid-nineteenth-century England were silent, submissive, and subservient. Not Lady Jane. She had publicly berated the Admiralty when they offered the £10,000 reward to the Hudson's Bay Company man, Dr. John Rae, for bringing back reliable information about Franklin's fate. Yes, Lady Franklin agreed, Rae had spoken to the Pelly Bay Inuit and retrieved stories about starving white men dropping in their tracks. He had traded trinkets for relics that proved these dying men were the same men who embarked on Franklin's voyage. But where, then, were the *Erebus* and the *Terror*? And where were the men, or the bodies of the men? Lady Franklin begged the Admiralty to search the shores of King William Island for more evidence of her husband and his likely demise.

In 1857 she made an impassioned plea to the Prime Minister, Lord Palmerston, begging that a final and exhaustive search be made on behalf of the "first and only martyrs to Arctic discovery in modern times." She had by then reconciled herself to the overwhelming likelihood that her husband was dead. But the government was deeply embroiled in the Crimean War. They politely but firmly turned her down.

Francis Leopold M'Clintock, a naval officer who had been involved in three previous Franklin search expeditions, could not do the same. He sympathized deeply with Lady Franklin. Moreover, he was not enamoured of the Admiralty's rigid adherence to procedure and protocol on search missions. In his previous voyages, M'Clintock had witnessed able-bodied seamen pulling heavy sledges across the ice while officers like himself stood aloof. Neither officers nor seamen were dressed for the climate, and the British seemed to have an aversion to living off the land. As a result, sledges were over-packed with goods that sweating and shivering men were forced to haul great distances in order to keep up some foolish appearance of civilization. Who was the navy trying to impress, M'Clintock wondered? The Inuit?

When Lady Jane's request arrived, M'Clintock leapt at the chance to implement a new search method, and he agreed to pilot the *Fox* to King William Island and the estuary of the Great Fish River as long as he could eschew the practices of the proper naval officer. He wanted to test his design for lighter sleds and equip his men properly for the climate. He would take his cue from the Inuit who populated the shores where, in M'Clintock's opinion, the bones of Franklin lay.

On July 2, 1857, ten years after the search had first been taken up, the refitted and reinforced 177-ton *Fox* sailed out of Aberdeen, Scotland, with a crew of twenty-five and provisions for twenty-eight months. Much of the supplies had been donated by a sheepish and repentant Admiralty.

Captain M'Clintock, who had taken a leave of absence from his naval duties, was sailing with Lieutenant William Robert Hobson as his second-in-command, and the company embarked in high spirits. They intended to follow up all the clues as to the whereabouts of the lost party in and around King William Island. Small sections of the coastline remained unexplored, and M'Clintock and Hobson planned to probe these and at last lay to rest the ghosts of Franklin's expedition.

Their hopes were soon deflated when, in the first sailing season, they became trapped in the pack ice of Baffin Bay and spent their first winter immobilized. With the little yacht frozen in, the pack ice moved south with the wind into Davis Strait. For the men on the ship, the movement would have been all but imperceptible, for the landscape was endless ice and snow, from horizon to horizon.

M'Clintock kept a journal of the voyage, and wrote a

haunting account of the mid-winter burial of a crewman named Scott, who fell down a hatch:

> *The greater part of the Church Service was read on board under the shelter of the housing; the body was then placed upon a sledge, and drawn by the mess mates of the deceased to a short distance from the ship where a hole through the ice had been cut; it was then "committed to the deep." What a scene it was! The lonely* Fox, *almost buried in snow, completely isolated from the habitable world, her colours half-mast high, and bell mournfully tolling; our little procession slowly marching over the rough surface of the frozen sea guided by lanterns and direction posts, amid the dark and dreary depth of Arctic winter, the deathlike stillness, the intense cold, and the threatening aspect of a murky overcast sky.*

In late April 1858, the *Fox* at last freed itself from the ice, and M'Clintock carried on north of the Baffin ice field to the entrance of Lancaster Sound. Through Parry's "Portal to the Arctic Archipelago" they sailed, arriving mid-August on Beechey Island, the now-notorious death site of Franklin's first three men, and the last place that held proof of his passage. Pressing on, M'Clintock's plan was to penetrate Barrow Strait and sail down the western side of Somerset Island, following the route he believed Franklin must have taken in 1846. The ice held him back, however, and a southern tack took the *Fox* down Prince Regent Inlet, following in the footprints of Ross and Parry.

Intent on threading the narrow channel of Bellot Strait,

identified six years earlier as the constricted waterway that divided Boothia Peninsula from Somerset Island, M'Clintock found winter had beaten him once again. He had no choice but to take harbour on the eastern side of Boothia Peninsula, at the entrance to the narrow strait. There, he and his men passed their second winter, preparing to set off to King William Island on foot as soon as the weather warmed sufficiently.

On his first overland journey, M'Clintock met with a group of Inuit near the Magnetic North Pole. They had in their possession many items from British ships. A brass naval button on the tunic of a sealskin robe was the first item that caught M'Clintock's attention. When he spoke to the people, a group of about twelve, they freely admitted that their relics were from a ship that had been crushed in the ice west of King William Island a few years ago. The men, they said, had come ashore and walked south. There were many of them, but it was many seasons ago. What happened to them was uncertain.

M'Clintock, encouraged, divided his search party into two groups: Hobson would scour the west coast of the island while M'Clintock followed the eastern shore toward the mouth of the Great Fish River, later renamed the Back River. M'Clintock would rendezvous with Hobson's party when his own party returned via the western coast.

It was early April 1859 when the two parties began their separate treks to circumnavigate King William Island and explore the mainland beyond. Without exception, the

Native people they met all told a similar story. One old woman said she remembered white men had walked to the south and "fallen down dead as they went." Another elder responded to M'Clintock's request for information by showing him silver utensils, knives, and spoons he had taken from a "tall boat with three arms" frozen into the ice to the west. There were *Kabloona* (white men), but they had died. Yes, he was sure they had all starved to death.

Reaching the mainland and trudging inland as far as Montreal Island near the mouth of the Back River, M'Clintock and his party found other indications that at least a few of Franklin's men had made it that far south: a piece of a tin can and some scraps of metal, barely rusted in the semi-arid conditions of the barren land. Still, the searchers couldn't be sure. Was it possible some of the men had found refuge with the Inuit? Perhaps they had taken wives and were still living with the nomadic people. Could all 129 men have died?

The persistent tales of starvation and death were confirmed on May 25, 1859, when a man in M'Clintock's party stumbled on a nearly complete human skeleton on a gravel ridge on the southern shore of King William Island. The party was on its way back to an appointed rendezvous site with Hobson when the bleached bones were discovered.

M'Clintock crouched by the body, still clothed in naval garb, face-down in the gravel. It looked as if the poor soul had fallen in his tracks from hunger and exhaustion. There were a few personal items scattered nearby: a clothes brush, a notebook, a small pocket comb with a few taffy-coloured hairs clinging to the teeth. The notebook put a name to the man: Harry Peglar, a petty officer on the *Terror*. His poor penmanship revealed little about the circumstances that claimed his life, but was devoted instead to general meditations on death. A passage of scripture could be deciphered amid the scrawl, but that was later revealed to be the handwriting of a different man, perhaps a quick epitaph written by a mate after Peglar stumbled and no longer had the strength to rise. The discovery extinguished the last hope that any member of the Franklin expedition remained alive.

William Hobson, meanwhile, was impatiently awaiting a reunion with his commander on the northwest coast of the island, for he had an even more startling manuscript to share. In the cleft of land that had been named Victory Point by James Ross in 1830 was a cairn with two messages scribbled on a single sheet of naval paper. The first, dated May 28, 1847, was encouraging, as Lieutenant Graham Gore of the Franklin voyage detailed the expedition to that date. The message, in Commander James Fitzjames's handwriting, read:

> *28 of May 1847. HM Ships* Erebus *and* Terror *wintered in the ice in Lat. 70° 05' N. Long. 98° 23' W. Having wintered in 1846–7 at Beechey Island, in lat 74° 43' 28" N, long 91° 39' 15" W, after having ascended Wellington Channel to Lat. 77°, and returned to the west coast of Cornwallis Island. Sir John Franklin commanding the expedition. All well. Party*

consisting of 2 officers and 6 men left the ships on Monday 24th May 1847.

Gm. Gore, Lieut. / Chas. F. Des Voeux, mate

Around the border of that encouraging account of the voyage was a message that had been written a full year later by Captain Francis Crozier, also dictated to and written by Commander Fitzjames:

April 25th, 1848—HM ships Terror *and* Erebus *were deserted on 22nd April, 5 leagues N. N. W. of this, having been beset since 12th September 1846. The Officers and crews, consisting of 105 souls, under the command of Captain F. R. M. Crozier, landed here in Lat. 69° 37' 42" N, long. 98° 41' W. This paper was found by Lt. Irving under the cairn supposed to have been built by Sir James Ross in 1831, 4 miles to the northward, where it had been deposited by the late Commander Gore in June, 1847. Sir James Ross' pillar has not, however, been found, and the paper has been transferred to this position, which is that in which Sir James Ross' pillar was erected. Sir John Franklin died on 11th June 1847; and the total loss by death in the Expedition has been to this date 9 officers and 15 men.*

James Fitzjames, Captain HMS Erebus
F.R.M. Crozier, Captain and Senior Officer

An important postscript read:

and start on tomorrow, 26th, for Back's Fish River.

Now they knew that Franklin was dead. And so, too, were those "105 souls" who had attempted to walk to "Back's Fish River" under Crozier's leadership. Hobson and M'Clintock mulled over the messages. There seemed one glaring error: in Gore's note he said the *Erebus* and the *Terror* had over-wintered on Beechey Island in 1846–47. Yet Franklin died in June 1847. Surely Gore had his dates wrong. The time when "all was well" must have been during the first winter 1845–46, so the vessels had been trapped in the ice north of Victory Point for nineteen months. At the time Fitzjames and Crozier had decided to lead the survivors away, April 26, 1848, twenty-four men, including Sir John, were already dead.

Yet, had the ships not been provisioned for three years when they sailed in May 1845? Franklin himself boasted they could likely last seven years with the supplies they carried. But according to Crozier's notation, it was a mere three years after their departure and the men were starving to death. It didn't make sense.

Just days later M'Clintock found a boat mounted on a sledge some fifty kilometres south of Victory Point. Curled up inside the boat, loaded guns at their sides, were two human skeletons, one partially consumed by animals, the other intact. This discovery raised more questions. The sledge and boat, a combined weight of nearly five hundred kilograms, were heaped with items M'Clintock deemed "a mere accumulation of dead weight, of little use, and very likely to break down the strength of the sledge-crews." The rescue party found clothing, crockery, books, boots, and toiletries, including scented soaps and silk handkerchiefs. Stranger still was the fact that the boat's bow was pointing north, back toward the trapped *Erebus* and *Terror*.

The only food M'Clintock's party could find was dried tea and twenty kilos of chocolate, leading him to conclude that the dead men at the site had been left behind as others staggered back to the ship in hopes of finding food. Perhaps the party had split, and some of the 105 had carried on to the south to meet the same fate as poor Peglar, whom they had discovered earlier. M'Clintock christened the land that cradled the frozen craft Cape Crozier, marked the details on his chart, and gathered his men for the trek back to the *Fox*.

Unlike so many before him, M'Clintock was fortunate. The ice on the eastern shore of King William Island was breaking up, and Prince Regent Inlet around Bellot Strait was nearly open water. The *Fox* was free to sail back to England in August 1859.

As his little ship traversed the northern seas, M'Clintock pictured the Franklin expedition sailing from their wintering grounds on Beechey Island south through an opening of clear water. Down they would have gone in the summer of 1846, through Peel Sound between Prince of Wales Island and Somerset Island until they reached the tip of King William Island. There Franklin would have looked at his charts and, believing Ross's assertion that a vast bay lay to the east, would have steered west into the polar ice floe that seized his boats forever on September 12, 1846. Like

M'Clintock himself two short winters ago, Franklin and his men would have waited patiently for spring. When it arrived, Gore was dispatched to the mainland, where he placed his note under a cairn at Victory Point when he reached shore on May 28. It was at some unknown date after that that he found things not well at all.

It would have been the following spring, May 1848, that the *Erebus* and the *Terror* were abandoned. After amending Gore's note, Croizer and Fitzjames struggled south with their much-weakened company of 105 men. What then? The Inuit said they had died. Some said they had taken to eating each other. M'Clintock shuddered. He did not know.

As the *Fox* rounded Cape Farewell and set her course for England, M'Clintock decided that the facts he had could only bring peace to Lady Jane Franklin. For eleven and a half years she had lived with the unknown. With certain knowledge that her husband was dead, her mind could have some rest, her heart some peace. But there was something still plaguing M'Clintock. If Franklin had chosen the different course, ignoring the charts that showed King William Island as a projection of the mainland, and sailed to the east of the island, would he not have been able to sail from sea to sea? The Northwest Passage was now certain in M'Clintock's mind. The channel to the east of King William Island was navigable, he was sure, and from there it was a simple matter of following the coastal waters and avoiding the polar ice to the north. Any sailor worth his salt could do it.

Yet M'Clintock knew he would not be the one to

attempt it. The frozen corpses he had seen on King William Island were men like himself who had felt the thrill of conquest as they attempted the Northwest Passage. Their dreadful fate had been etched forever in his mind. After reporting his discoveries to the navy and to the woman who had commissioned his trip, he put aside his vision of a winding passage that would take a ship from Baffin Island to the Bering Sea. Franklin was lost and the Northwest Passage found, but for Francis Leopold M'Clintock, the opposite was true.

ROALD AMUNDSEN

Through the Passage

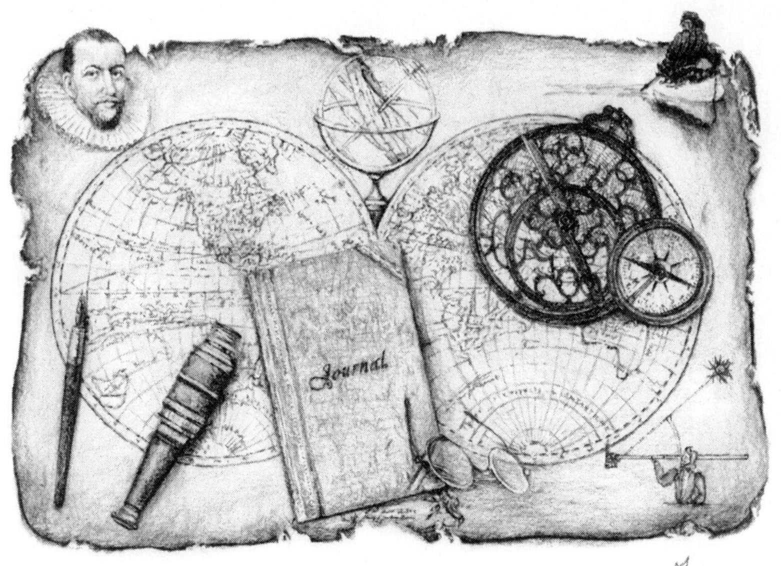

The 350-year quest to find and sail the Northwest Passage finally ended in the first decade of the twentieth century when a slim Norwegian named Roald Amundsen took six friends and a forty-seven-ton, single-masted sloop and sailed through the tortuous channel that had defeated hundreds of Arctic explorers before him. The vessel, the shallow-draft, twenty-two-metre *Gjoa,* was less than a quarter the size of the ship Martin Frobisher had been commissioned to load with Arctic gold in 1577. But it was no ordinary boat. And Amundsen was no ordinary man. First, he had amazing luck in terms of the unpredictable Arctic weather. Second, he had the determination to follow a dream he had nurtured from the time he was a child haunting the docks of Christiania, the city now known as Oslo.

After the mysteries of the Franklin expedition had been more or less resolved, the British drew away from Arctic exploration. From M'Clintock's departure in 1859 until 1880, it was two Americans who raised the banner. Charles Francis Hall, a Cincinnati businessman, and, after him, Lieutenant Frederick Schwatka, an officer of the American cavalry, gathered oral histories of cannibalism from the Inuit, scoured grave sites and death camps, and gradually pieced together what M'Clintock had already surmised.

It was Schwatka who tracked the relics and rumours of the ever-diminishing Franklin expedition across Simpson

Strait to the Adelaide Peninsula, and he who named that final resting ground Starvation Cove. The last survivors of the *Terror* and the *Erebus* had perished there, he reasoned, either unable or no longer willing to consume the human limbs and bones they had dragged with them in a box. While little evidence was found at Schwatka's site, one partial body was recovered. According to the Inuit of the region, the remains of the others, presumed to number anywhere from twenty to thirty men, had disappeared forever beneath the shifting sands of the northern desert.

In 1880 Britain transferred sovereignty of the Arctic to Canada. The human costs of their Arctic pursuits may have lain heavily on the nation's conscience. In any case, a new contest had captured the attention of the world: the race to reach the North and South Poles. Armed with knowledge of what wind and ice can do to a ship, the contest had become an international, overland foot race to the top and bottom of the globe.

The Norwegians came into their own during the last two decades of the 1800s. While Roald Amundsen was immersing himself in the exploits of Franklin, Scandinavian men were making Arctic history. Norwegian scientist Fridtjof Nansen spent three years drifting across the polar basin in his specially designed craft, the *Fram*, while Swedish adventurer Salomon August Andree perished trying to reach the North Pole in an ill-fated ballooning expedition. Another Norseman, Otto Sverdrup, refitted Nansen's vessel in 1896 and, aided by huskies and skis, explored and mapped the entire northwestern side of Ellesmere Island.

Although Norway was still a province of Sweden, the Norwegians were leaving their distinctive mark on the Arctic with a view to gaining independence in the near future. A movement was afoot to declare Norway independent by 1905.

Roald Amundsen was reared by the sea, and his imagination had been shaped by seafaring people. When the *Fram* returned to Christiania from her first stint on the Greenland ice shelves in 1889, the seventeen-year-old Amundsen was there in the cheering crowds that greeted the ship. That could well have been the event that confirmed in him the desire to sail the Northwest Passage. Steeped in tales of British heroism and suffering, and caught up in nationalistic excitement as his own countrymen took up the challenges of the north, it was the natural next step in the evolution of his character. While the Northwest Passage had been charted and even walked, no one had actually sailed its length. Roald Amundsen decided that he would be the one to do it.

After he had made the decision, there was no stopping him. He quit medical school to begin a study of polar navigation. He underwent vigorous training, to the extent of signing on for two seasons on a polar sealing vessel to accustom himself to the extreme conditions of the north. Hours of cross-country skiing were broken only by equally long hours of study about the Canadian north.

Like M'Clintock and Parry before him, Amundsen felt that his best chance of navigating the passage was to approach it from the east and sail as close to the shore as possible. Sailing the channels of water that opened between the mainland and the offshore islands looked to be a far more likely possibility than fighting through the thick of the polar ice streams. But the channels might prove shallow. For this reason, Amundsen purchased the thirty-one-year-old *Gjoa,* a refitted fishing boat, feather light and shallow in the water.

Amundsen pored over accounts of previous expeditions. Knowing his journey would take more than one season and already in debt from the purchase of his ship, he went in search of a backer. But who would outfit a ship the size of the *Gjoa*? And who would pay Amundsen and his mates a living wage without some guarantee of information or recognition in exchange? Amundsen decided the expedition would have to have a scientific component if it were to receive any reasonable financing.

Fortunately, a debate about whether the Magnetic North Pole, identified by James Ross, was static or shifting was ongoing in the European scientific community. Amundsen would go to the area described by Ross and register magnetic recordings, thus proving that his journey had scientific intent. If men of science wanted to believe the exact location of the Magnetic North Pole was more important than the conquest of the Northwest Passage, let them. It was a fine solution, and one that piqued the interest of the German Marine Observatory in Hamburg. When Amundsen pitched the idea, scientist and scholar Dr. Geheimrat Georg von Neumayer responded with enthusiasm. It was a near-perfect match. Von Neumayer supplied the funding and Amundsen and his carefully selected crew of six supplied the brawn, the brains, and the wherewithal to complete the study.

On June 16, 1903, Amundsen and his companions departed from Christiania. The sail around Greenland took upwards of two months, but Amundsen had chosen his crew well. All but two of the six had had previous polar experience. A zoologist, an engineer, a meteorologist, a couple of experienced sailing hands, and a man proficient with a harpoon made up the youthful and energetic team. Most important, all had the same mandate: to sail the *Gjoa* through the Northwest Passage or sacrifice themselves in the attempt. There were no doubters amongst them. Indeed, while the little ship was thrashed about by wind and sea, the crew took it in stride, and seemed even to embrace the dangers around them.

"There is the sound, dead ahead, on our lee side," second mate Helmer Hansen called from his watch in the crow's nest on August 20, 1903. As the vessel swung into the wind and tacked west, Lancaster Sound opened out before them. Two days later, the *Gjoa* anchored on Beechey Island. Amundsen recorded his thoughts in a journal he kept throughout the voyage:

It was about 10 o'clock when twilight came on. I was sitting on one of the chain lockers looking towards the land, with a deep, solemn feeling that I was on holy ground: Franklin's last safe winter harbour. My thoughts wandered back—far back. I pictured to myself the splendidly equipped Franklin expedition . . . officers in dazzling uniforms, boatswains with their pipes, blue-clad sailors; proud representatives of the world's first seafaring nation up here in the unknown ice waste! At the breaking up of the ice in 1846, the Erebus *and the* Terror *again stood out to sea . . . once more waves England's proud flag; it is the farewell of the Franklin expedition. From this point it passed into darkness—and death.*

The tenor of the Norwegian trip seemed to change after this rumination on the doomed Franklin. The *Gjoa* was now pressing south into lesser-known waters, and although ice conditions were unusually favourable, Amundsen was acutely aware that others had suffered and died in the same waters he now cruised with ease.

They sailed between Prince of Wales Island and northern Somerset Island, to Prescott Island in Franklin Strait, where their compass needle became useless. From that point on the young Norwegians were dependent on celestial navigation and a handful of dubious charts. Heavy fog descended as they made their way down Franklin Strait. At one point they heard the dreaded sound of a keel scraping against rock, and the men scrambled to lighten their cargo. They had grounded amidships, and despite setting all the sails, starting their small motor, and throwing out a kedge anchor, a number of items—immediately dubbed non-essentials—had to be cast into the sea in order to float the *Gjoa* free.

At the top of King William Island, where Franklin made his fateful error, Amundsen steered southeast, to the inland passage between the Boothia Peninsula and the windswept shores of the island. Although not ice-free, the channel Ross had mistakenly called an inlet was fairly simple to navigate, and by the beginning of September the *Gjoa* and her crew were searching for a place to harbour for the winter.

Their luck held. On September 12, as the sky descended and the days continued shorter and briefer, the *Gjoa* sailed into a tidy cove on the southeast corner of King William Island. Named Gjoahavn, now a hamlet of some six hundred people known as Gjoa Haven, the harbour was nicely protected from the wind and proved to be an excellent wintering ground. Caribou were plentiful, and as Amundsen's men established a base camp on shore they took advantage of the migrating herd and stocked their ship with fresh meat.

It was while the crew were draping the ship with sailcloth in preparation for the winter that they encountered a band of Inuit. They were first sighted on the crest of a hill, descending toward the ship. Amundsen was clearly nervous. He had read many accounts of European encounters with Inuit, and he was not sure what to expect.

"I'll go, but I'll go armed," he said to his comrades, hiding a pistol in his belt. A second man, Godfred Hansen, chose to accompany the captain, and as the two trudged across the ice, each laid a hand against his weapon. It was an unnecessary precaution. The Inuit seemed delighted to see men and a ship. It was as if a legend told by their elders had sprung to life. Two generations had passed since the story of James Ross had circulated among this particular community. Without knowledge of Franklin and his various pursuers, Ross would have been the only European they would have heard about. Now, more then seventy years after Ross had struggled to the Magnetic North Pole, the bearded strangers were back. The Inuit threw down their own weapons and moved to embrace them. Amundsen and Hansen breathed a sigh of relief. Relations, it seemed, would be amicable.

The Inuit soon set up a camp near the frozen ship, and the two races initiated a barter system. In exchange for showing the white men how to make robes and mukluks from caribou hides, the Inuit learned the art of cross-country skiing. A soapstone, seal-blubber lamp was exchanged for six iron needles and a small metal grill. A pocket mirror was given to an Inuit man who, in return, taught the foreigners how to build an igloo from blocks of packed snow.

The winter of 1903–04 was fruitful and generally cheerful. The Natives kept the Norwegians entertained and taught them a good deal about Arctic survival. For their part, the Europeans came to love and respect the good-natured people of the north. Amundsen wrote in his journal:

> *I must state it is my firm conviction that . . . the Eskimo, living absolutely isolated from civilization of any kind, are undoubtedly the*

happiest, healthiest, most honourable and most contented [people]. It must, therefore, be the bounden duty of civilized nations who come into contact with the Eskimo, to safeguard them against contaminating influences, and by laws and stringent regulations protect them against the many perils and evils of so called civilization . . .

The arrival of the New Year saw the sailors keen to begin the scientific side of their voyage. Gustav Wiik, the meteorologist in the party, set up an elaborate system of observations from a crest named Magnetic Hill, and the self-registering instruments recorded data that would determine the static nature of the Magnetic North Pole. A number of trips to the charted position of the pole were attempted, but temperatures dipping to –79°F caused a hasty retreat. On April 6 the cold snap broke and Amundsen, Hansen, and first engineer Pedar Ristvedt travelled on foot to Boothia Peninsula, where they discovered that the site Ross had pinpointed as the Magnetic North Pole on June 1, 1831, had moved some forty-eight kilometres northeast. The pole was not stationary, a discovery that confirmed Wiik's magnetic observations.

It was a cold spring. The summer of 1904 found the *Gjoa* still ice-bound and the Northwest Passage unconquered. A second season with the Inuit was inevitable, so the seven Scandinavians settled down to make the most of it. Adolf Lindstrom, the company zoologist, was recording fauna in the area, and was astounded at the variety and abundance of wildlife. On an expedition south to Hunger

Bay, so named because of the discovery there of a number of skeletons from Franklin's expedition, Amundsen ruminated on the different experiences:

> *In spring , when the channels are opened, enormous quantities of large fat salmon are met with. A little later the reindeer arrived in countless hordes and remained here throughout the summer, then in the autumn an unlimited quantity of cod can be caught, and yet here—in this Arctic Eden—those brave travellers died of hunger. The truth probably is that they had arrived there when the low land was covered over with snow; overcome by exertions, worn out with sickness, they must have stopped here and seen for miles before them the same disheartening snow-bedecked lowland, where there was no sign indicating the existence of any life, much less riches, where not a living soul met them to cheer them up or give them encouragement and help. Probably there is not another place in the world so abandoned and bare as this in winter.*

On November 20, the sailors got a surprise visit from an English-speaking Inuit, Mr. Atangala, who had heard about the ship on King William Island while paddling the Coppermine River. He was keen to use his English skills to assist the Norwegians, and offered to travel to Hudson Bay where he knew a Canadian ship, the *Arctic,* was exploring the northern reaches of the bay. Amundsen was keen to establish contact with the outside world and agreed to let Atangala act as a long-distance messenger. He departed for the south with lengthy letters penned by the crew strapped to a pouch on the side of his parka. His destination was Roe's Welcome Sound where the *Arctic* was over-wintering at Cape Fullerton. It was a journey of some 750 kilometres, and they thought they had seen the last of him. To their utter surprise, Mr. Atangala reappeared on Saturday, May 20, out of a bitter northeasterly gale with a soldered tin box containing a packet of return post strapped to his sled.

"It's mail," cried a jubilant Amundsen, throwing his arms spontaneously around the diminutive Inuit. "What news? What news?" and he scrambled for the box, surrounded by the others.

While the box did not contain personal missives from kith and kin, it did have a lengthy letter from a Major Moodie, the chief of the Northwest Mounted Police aboard the *Arctic*. A second letter from the ship's captain alerted Amundsen to American whaling vessels on the northwest coast of the continent. The news sent a visible shiver down the young Norwegian's spine.

"We're close," he whispered to his shipmates. "We are so close to being the first through."

Like his ship-bound colleagues, Amundsen devoured the newspaper clippings that were included in their package, but no amount of international news could shake from his mind the image of whaling vessels to the west. When the *Gjoa* reached them, the whole world would know that they had done what no one had ever done before.

His contentment with the camp deteriorated markedly after the post arrived. Amundsen, again besotted with his dream, was becoming increasingly anxious to flee Gjoahavn.

A cool May hastened the warming month of June, and 1905 promised to be equal to the glorious summer of 1903. In July the melt was thorough and quick, and by August 1 the ice was off their harbour and Rae Strait was a promising shade of blue. The *Gjoa* was rolling with a northeasterly wind. On August 13, they cast off into the virgin waters of the Northwest Passage.

From Booth Point through the narrow gutter of Simpson Strait, the *Gjoa* navigated waters that ran from seventeen to five fathoms in the blink of an eye. Ice competed with rocks to destroy the sloop. After two years ashore, the sailors got a quick lesson in liveliness and the importance of keeping a sharp eye out. They slept little on the treacherous journey around the southern tip of King William Island. The narrow channel, though ice-free, was strewn with rocks and reefs, and theirs was a chaotic, zigzag course that only just managed to avoid disaster on several occasions. Then, a mere four days after departing Gjoahavn, they sighted Cape Colbourne on the southeast tip of Victoria Island.

They had done it. Roald Amundsen and his crew had joined east to west. The Northwest Passage was won!

For three weeks they sailed the waters of Queen Maud Gulf between the mainland and Victoria Island until, on August 26, 1905, they met with a second ship. Amundsen was sleeping when the word came down from the crow's nest: "Vessel in sight!"

He bolted from his bed and leaped onto the deck. To

the northeast was an American whaler. Amundsen felt his throat constrict; his heart seemed to swell in his chest. As the whaler came closer, tears of relief and wonderment sprang to his eyes. The childhood dream was fulfilled—and not just his childhood dream, but the collective dream of all his heroes: Frobisher, Hudson, Foxe, James, Ross and Parry, Franklin, and those who searched for him, Rae and McClure and M'Clintock and Hall. All of them.

It was a defining moment. When the two ships came alongside, the crew of the *Gjoa* could barely speak. Yes, they had come from the east. Yes, the Northwest Passage had now—just now—been traversed.

The day—August 26, 1905—was written on their hearts, and would soon be committed to history for all to see. Roald Amundsen, a thirty-three-year-old Norwegian, and his youthful team, had made the dream of the Northwest Passage a reality.

As though to remind them of the hardships of those who had gone before, the Arctic weather promptly wrapped itself around the *Gjoa* and sealed Amundsen's ship in the ice for one last winter. By the end of September the ship was immobile at King's Point, just west of the Mackenzie River delta.

Amundsen was like a caged animal that winter, so anxious was he to broadcast the news of his triumph. Finally, in distraction, he decided to team up with a band of Inuit hunters and a stranded American whaling captain, William Mogg, for an eight-hundred-kilometre trek overland to the nearest telegraph office in Eagle City, Alaska. It was there, December 5, on the banks of the Yukon River, with the temperature hovering at –60°F, that Roald Amundsen, the first navigator of the Northwest Passage, telegraphed his news to his countryman, mentor, and fellow northern explorer, Fridtjof Nansen. The message read, in part:

> *We did it!*

The arch at the top of the world was closed in the summer of 1906 when the *Gjoa,* steered by Amundsen, rounded Point Barrow, Alaska, and headed south through the tranquil waters of the Pacific. A ship had finally sailed through the Northwest Passage from coast to coast. China and the riches of the Orient were no longer the pull; the notion of finding a trade route had long since been supplanted by ambitions of science, conquest, national pride, and personal glory. The quest that had stolen so many lives and encouraged countless acts of foolishness and heroism was at last fulfilled. As Amundsen and his men sailed down the western coast of North America to the port of San Francisco, they knew theirs was a shared victory.

Epilogue

The Northwest Passage is still there, a fleeting crack in a frozen landscape that opens up once in a while to allow a few vessels to pass through its awesome icefields.

In the past decade, global warming and a depleted ozone layer over the Arctic have had their way with the passage. There is now the very real possibility that ice breakers, tankers, trawlers, and the mammoth vessels that carry the world's natural resources could use the Northwest Passage as a regular shipping route. It might even become the trade channel between Europe and the Far East that was dreamed of in the sixteenth century.

The Northwest Passage of tomorrow will not carry the coveted materials of Frobisher's day—silk, spices, and gold—but rather timber, water, and oil, the precious raw materials and commodities that over time will become scarcer and scarcer in an age of voracious consumption.

The notion of a clear shipping passage across the top of North America, then, comes with a cautionary message, one that would have to be weighed against the environmental and cultural impact of such a route on a fragile ecosystem. Questions of national defence, sovereignty, and Native interests arise unbidden out of the future plans for the Northwest Passage.

And because history so often repeats itself, it would be wise to reflect on the initial opening of the Arctic waters. The Northwest Passage was won by men of dubious vision, who were willing to sacrifice all for the opportunity to engage in a battle against the elements, against time, and against the frailties of the human body.

Many died in that strange and ferocious quest, for the Arctic often claimed the little lives that danced briefly on its shores. More often, it spat them back toward Europe, whence they came. They were a special breed of European adventurer who could not let a landscape triumph over them, yet found it difficult to adapt to a way of life that had been practised by the people of the north since the beginning of time. Yet slowly, cumulatively, painfully, over 350 years, they won a way between the granite shore and the perpetual ice and claimed the Northwest Passage for the world.

To what end?

The Northwest Passage was conquered because it is there, because it could be done, and because, above all, it ignited in a few human hearts the spirit that is not afraid to engage itself against fearsome odds. Now that we know it—now that the Arctic islands are charted and placed upon our maps with accurate coastlines and official communities—it is difficult to imagine what it must have felt like for those first men to strike out into the unknown.

These men of the past—Frobisher, Davis, Weymouth, Hudson, Munk, Bylot, Baffin, Foxe, James, Knight, Ross, Parry, Franklin, McClure, Belcher, M'Clintock, Hall, and Amundsen—can be counted among those who gave the people of today and tomorrow the possibilities the Northwest Passage still holds.

Chronology

1497–98	John Cabot claims Newfoundland for the English and maps eastern coast of North America in two voyages.
1534–42	Jacques Cartier makes three voyages of exploration to North America, surveying the coast of Canada and the St. Lawrence River, and providing the basis for later French claims in the area.
1558	Elizabeth I becomes queen of England. England's sea power stimulates voyages of discovery during her reign (1558–1603).
1576–78	Martin Frobisher undertakes three voyages to Labrador and Hudson Bay and reaches Baffin Island. Frobisher Bay is named after him.
1584	English navigator Walter Raleigh sends the first of three expeditions to North America.
1585–87	John Davis commands three voyages in an attempt to find the Northwest Passage. Discovers Davis Strait.
1602	George Weymouth begins trip through Hudson Strait but is forced back by ice.
1607	Henry Hudson searches the Barents Sea for the Northwest Passage.
1608	Samuel de Champlain founds Quebec.

1609	Hudson begins expedition of the northeast coast of North America. Sails into Hudson Bay.
1610	Hudson's fourth voyage in search of the Northwest Passage. He explores the strait and bay that will bear his name.
1611	Hudson's crew mutinies and abandons Hudson, his son, and seven others in a small boat. They are never heard from again.
1612–16	William Baffin pilots several expeditions in search of the Northwest Passage. Under the command of Robert Bylot, visits Hudson Strait in 1615 and are the first Europeans to find Baffin Bay (1616).
1620	Jens Munk and two crewmen are the only survivors from a crew of sixty-four after wintering on the shores of Hudson Bay in 1619–20.
1631	Luke Foxe explores the west and northwest of Hudson Bay. Thomas James also sails in search of the Northwest Passage.
1759	James Wolfe, British major-general, commands the British victory at Quebec.
1818	John Ross and William Parry follow Davis Strait through Baffin Bay.
1819–20	William Parry's first voyage in search of the Northwest Passage. His expedition penetrates the Arctic Archipelago and reaches Melville Island.

1819–22 John Franklin's first overland expedition to Point Turnagain in search of the Northwest Passage. Eleven men lose their lives.

1821–23 William Parry's second voyage in search of the Northwest Passage. Reaches Fury and Hecla Strait.

1824–25 William Parry's third and final voyage in search of the Northwest Passage.

1825–27 John Franklin's second overland expedition. Maps more than 1600 kilometres of coastline from Coronation Gulf to Icy Cape, Alaska.

1829–33 John Ross's second expedition in search of the Northwest Passage. Ross and his crew spend four winters in the Arctic.

1831 John Ross raises the Union Jack at the Magnetic North Pole.

1841 John Ross raises the Union Jack at the Magnetic South Pole.

1845–47 Sir John Franklin's expedition is last seen by European eyes in 1845, north of Baffin Bay. All 129 hands are lost in the Arctic.

1848–49 James Ross leads one of many expeditions that will be mounted over the years by various parties in search of Sir John Franklin.

[illegible] Robert McClure leads expedition through the Bering Strait in search of Franklin and confirms the existence of a Northwest Passage.

1857–59 Francis Leopold M'Clintock leads expedition that confirms Franklin's fate.

1880 British sovereignty over the Arctic passes to Canada.

1903–06 Roald Amundsen and his crew sail the Northwest Passage for the first time in history.

References

Beattie, Owen and Geiger, John. *Frozen in Time: Unlocking the Secrets of the Franklin Expedition.* Saskatoon, SK: Western Producer Prairie Books, 1987.

Burton, Pierre. *The Artic Grail: The Quest for the Northwest Passage and the North Pole 1818–1909.* Toronto, ON: McClelland & Stewart, 1998.

Delgado, James P. *Across the Top of the World: The Quest for the Northwest Passage.* Vancouver, BC: Douglas & MacIntyre, 1999.

Francis, Daniel. *Discovery of the North: The Exploration of Canada's Arctic.* Edmonton, AB: Hurtig Publishing Ltd., 1986.

Hakluyt, Richard. *The Prinipall Navigations, Voiages and Discoveries of the English Nation.* London, Eng.: Bishop and Newberie, 1589.

Hiscock, Eric. *Cruising Under Sail.* Oxford, Eng.: Oxford University Press, 1981.

Keating, Bern. *The Northwest Passage from Matthew to the Manhattan 1497–1969.* New York, NY: Rand & McNally & Co., 1970.

M'Clintock, Francis Leopold. *The Voyage of the Fox in the Arctic Seas.* London, Eng., 1860.

Mowat, Farley. *Ordeal by Ice.* Toronto, ON: McClelland & Stewart, 1973.

Neatby, Leslie H. *In Quest of the Northwest Passage.* Toronto, ON: Longmans, Green and Co., 1958.

…in. *Narrative of a Second Voyage in Search of the Northwest …ssage.* London, Eng.: A. W. Webster, 1835.

Ross, M. J. *Polar Pioneers: John Ross & James Clark Ross.* Montreal, PQ: McGill-Queen's University Press, 1994.

Rundall, T. *Narratives of Voyages Toward the North-West.* London: Hakluyt Society, 1849.

Sutton, Ann and Myron. *Journey into Ice: Sir John Franklin and the Northwest Passage.* New York, NY: Rand & McNally & Co., 1965.

Thomson, George Malcolm. *The Northwest Passage.* London, Eng.: W& J Mackay Ltd, 1975.

Woodman, David C. *Unravelling the Franklin Mystery: Inuit Testimony.* Montreal, PQ: McGill-Queen's University Press, 1991.